Communities and Alternatives:
An Exploration of
The Limits of Planning

In COMMUNITIES AND ALTERNATIVES, Shimon S. Gottschalk undertakes a sociological study of the responsiveness of communities to change. Challenging the utopian fantasy of the unchanging perfect society, Professor Gottschalk contends that all societies, future as well as present, must continually evolve in response to changing realities as well as utopian ideals. There must be a balance between underplanning and anarchy, and overplanning and totalitarianism.

Dr. Gottschalk is concerned with the paradox of planning the communal organization, which is intrinsically low goal-oriented and non-hierarchical. He defines three levels of communal society — external organization, community level and family level — and eight possible combinations of relationships between formal and communal organizations. Having laid a theoretical groundwork, Gottschalk uses his model to compare the different orientations of his three case communities of this book:

*Moosehaven, an administered retirement community

*Levittown, L.I., a designed suburban community

*The Bruderhof, an intentional religious community

The volume helps in providing the basis for the planning of humanistic communities. Writes Professor Roland Warren in his Preface to this work, "The implications for social planning of Gottschalk's analysis of the distinction between formal and communal organization should be read by everyone who seeks, through social planning, to correct or modify perceived imperfections in the current system."

# COMMUNITIES AND ALTERNATIVES

## AN EXPLORATION OF THE LIMITS OF PLANNING

SHIMON S. GOTTSCHALK

Schenkman Publishing Company, Inc.

Halsted Press Division
**JOHN WILEY AND SONS**
New York    London    Sydney    Toronto

Copyright © 1975
Schenkman Publishing Company, Inc.
Cambridge, Massachusetts 02138

Distributed solely by Halsted Press, a Division
of John Wiley & Sons, Inc. New York.

Library of Congress Cataloging in Publication Data

Gottschalk, Shimon S.
    Communities and alternatives

    Bibliography: p.
    1. Community. I. Title.
HT65.G68        309.2'6            74-22269
ISBN 0-470-31907-0
ISBN 0-470-31908-9 pbk.

Printed in the United States of America

To Takana' Ariela

# CONTENTS

Foreword                                                                    xiii

Acknowledgements                                                              xv

Chapter 1
  Introduction                                                                 1

Chapter 2
  Communal Organizations—A Summary                                            9

Chapter 3
  A Definition of Community                                                   18

Chapter 4
  A Classification of Community Types                                         25

Chapter 5
  The Measurement of Change on the Community Level                           34

Chapter 6
  Moosehaven: An Administered Community                                       44

Chapter 7
  Levittown, Long Island: A Designed Community                               58

Chapter 8
  The Bruderhof: An Intentional Community                                     71

Chapter 9
  Summing-Up                                                                  94

Chapter 10
   The Idea of Community                           100

Chapter 11
   The Limits of Planning                       106

Chapter 12
   Toward a Scenario for an Alternative Society     113

Glossary                                  130

Appendix A
   The Theory of Communal Organizations        133

Appendix B
   Community Schedule                       158

Selected Bibliography                     163

## LIST OF PLATES AND TABLES

Plate 1        The Similarities and Differences between Formal
               and Communal Organizations                          10

Plate 2        Two Meanings of the Word Community                   18

Plate 3        The Three Levels of Community                        19

Plate 4        A Classification of Community and Anti-Commu-
               nity Types                                           32

Plate 5        An Operational Classification of Community
               Types                                                35

Plate 6        The Dimension of Level II Goal Orientation in
               Communities                                          37

Plate 7        The Dimension of Staff-Resident Split in Com-
               munities                                             38

Plate 8        The Dimension of Decommunization                    39

Plate 9        Community As Sentiment in Communal and For-
               mal Organizations                                   101

Plate 10       The Translation of Community and Anti-Com-
               munity Types from Contemporary to Alternative
               Society                                              119

Plate 11       Societal Types and Their Normative Emphasis         124

Plate 12       The Primary Types of Cooperation Among the
               Elements of Organizations.                          139

Plate 13       The Linkages of Formal and Communal Organi-
               zations, Cities, and Folk Villages                  143

Table 1        Age Distribution of Population in Levittown,
               N.Y., 1950, 1960, 1970                               63

# FOREWORD

In the current obsessive proliferation of books dealing with "urban problems" it is seldom that one finds a work which opens up a new vein of theoretical understanding. Shimon Gottschalk's primary mode of analysis in this book has been theoretical. In pursuing it, he has examined in detail three different "deviant" communities, all of them small. Yet the book has important practical implications for communities both large and small.

What makes Gottschalk's work basic is his systematic conceptual analysis, carefully constructed, which puts to further use the important distinction between formal and communal organizations developed by George Hillery. It goes beyond Hillery's important analysis with an ingenious conceptual approach to exploring the relationships between the community level and the societal level on the one hand and the intracommmunity level on the other. The distinction between formal and communal organizations and the relationships among these three levels form the framework for an analysis which is truly architectonic. Concepts are used not idly, but as building blocks for understanding, as means of passing back and forth between the detail and the larger configuration, and as means of laying the theoretical basis from which important conclusions are derived for such community-related developments as new towns, communes, and the counter-culture. The implications for social planning of Gottschalk's analysis of the distinction between formal and communal organizations should be read by everyone who seeks, through social planning, to correct or modify perceived imperfections in the current situation. Likewise, his closing analysis of communal systems in which goal-orientation is normatively evaluated as negative rather than positive presents not only a useful analytic structure for understanding various aspects of the counter-culture, but a basis on which to assess critically the alternatives which it affords.

Since most recent work in the community field, including my own, has been addressed largely to formal organizations in their relationship to each other at the community level and to formal organizational structures beyond the borders of the community, Gottschalk's emphasis on communal organizations rather than formal organizations is especially to be welcomed. Quite understandably, in placing this emphasis, Gottschalk took Hillery's book on *Communal Organizations* as his point of departure, a work which has been, in my estimation, the most important contribution to community theory to appear in recent years.

Although Gottschalk's analysis is truly radical, one does it an injustice by applying to it such overworked terms as counter-culture, anarchism, non-violence, and the like. His book opens and closes with considerations which include all of these. Yet the bulk of his book does not take them as its point of departure, but quite the contrary, it builds a thoroughly dispassionate analytic structure which enables him to address these considerations from the standpoint of a carefully wrought intellectual edifice, rather than from the usual hortatory and largely journalistic tracts to which we all have become accustomed in recent years.

Alvin Gouldner has occasionally made the telling criticism of "radical" sociologists that their radicalism remains somewhat unrelated to their sociology, which is highly conventional. Gottschalk's work is not susceptible to this criticism; not because it is not radical, or because it is not conventional, but precisely because a highly skillful and creative application of conventional modes of analysis turns out to be directly related to a set of quite radical implications.

Roland L. Warren
Brandeis University

# ACKNOWLEDGMENTS

This study was germinated in a place far removed from the lecture halls and libraries of Brandeis University. I was engaged in a novel project: the effort to establish a new town for poor people, former sharecroppers and farm hands, in rural Georgia. The idea was—and continues to be—to develop another option, a viable alternative to urbanization, for black families that have been deprived of their livelihood. They have been disinherited largely as a result of U. S. agricultural policy, which has been consistently biased against small, poor, and especially black farmers.

But how does one plan, how does one create a new community? Building a new *town* is a relatively easy task, once the talent and the necessary material resources have been assembled. But how does one create a new *community?* This question pressed ever harder upon us, the self-styled experts, as we became increasingly sensitive to the fact that the people with whom and for whom we were attempting to plan were many leagues removed from us in culture, in life style, in knowledge, and in values.

As time went on and our project progressed only haltingly, we became intimately, even viscerally aware of our limitations as planners. That is why this volume is centrally concerned with the question of identifying and specifying insofar as possible, the area which should not be planned.

I am grateful to the people of Southwest Georgia for teaching me so many important lessons of life. I am thankful most especially to my good friend and co-worker Robert S. Swann, without whom New Communities, Inc., would never have been started. The late Slater King, Charles Sherrod, C. B. King, James Mays, Marion and Leonard Smith, Cellestine Hill, Mrs. Mack, Robert Christian, Robert McClary, Robert Gaines, Mr. L. B. Johnson, and countless others who have worked to create New Communities, Inc., have served as an endless source of insight and inspiration for a research task of whose existence they were only vaguely aware.

Of the many persons who contributed to this study, the most influential has been Roland L. Warren, professor of community theory at Brandeis University. His exceptional talent for carefully dissecting the logic of every argument, his knowledge and wisdom, his tolerance for ideas that deviate from his own, and most important, his endless encouragement, have been a source of irreplaceable support.

George A. Hillery, Jr., whose study of communal organizations has been the fountainhead of much of the theory developed in this study, has been an exceptionally devoted and patient teacher. Without his personal involvement and detailed criticism this study would have been vastly different.

David Austin has been a constant mentor and critic at my right hand. He has been a friend and guide, wonderfully available in times of need.

Rosabeth M. Kanter was especially helpful in strengthening the methodological aspects of this study. Her knowledge of American utopian communities of the nineteenth century and her great personal experience in community studies were of great value.

Without Marianne Muscato and Roberta Hobbs, who typed and retyped these pages at an almost incalculable speed, this study would never have progressed so quickly. My colleagues Richard Brooks and June Hopps are abundantly deserving of thanks. I am grateful to the staff of the Center for Community Economic Development for making available out of their budget essential travel funds during a period of acute personal financial need.

The families who served as our hosts at the Bruderhof, Mr. Charles McCall of Moosehaven, Dr. Will Justiss of Jacksonville, Mr. Dennis King and many other kind people in Levittown—all these and many more are deserving of thanks.

This study is dedicated to Takana' Ariela whose period of gestation coincided almost identically with the writing of this volume. To her mother, my wife Elizabeth, who cautiously mediated the rivalry between both babies, one in-and the other ex-utero, go expressions of thanks which are beyond words.

Brandeis University, and
Ashby, Mass.

**Communities and Alternatives:
An Exploration of
The Limits of Planning**

In the beginning, God created the world with a measure of Justice and a measure of Mercy. Had he created it only with Justice, it would have cracked; had he created it only with Mercy, it would have crumbled.

— from the Midrash

# CHAPTER 1

# INTRODUCTION

We are on the verge of a new era. Let us be blunt and call it by its proper name: The Age of Retrenchment. The rate of technological progress will begin to wind down, population control will begin to be effective, and eventually, a stable world population will be reached. The great urban centers will begin to disperse, and the Western world will again move in the direction of what Howard Becker called a sacred society.[1]

Planners, especially social planners, a loosely defined group of professionals with whom the author is frequently identified, are not supposed to be soothsayers or prophets, and the predictions which are made in these opening lines are not intended in such a vein. But by definition, planners are concerned with the future—with a knowledgeable and presumably rational assessment of that which is soon to be history. At the core of such an assessment is the judgement of what can, and what cannot, be controlled. That is what planning is all about: the attempt rationally to control future events, in the light of that which cannot be controlled.

Most of those who in recent years have concerned themselves with a projection of the future have exercised a variety of methods of extrapolation from the present. The "basic, long-term, multifold trend" of Western society is assumed to be inexorably in the direction of increasing industrialization, urbanization, bureaucratization, affluence, leisure,

---

[1] Howard Becker, *Social Thought from Lore to Science* (Washington: Harren Press, 1952) Vol. 1, pp. 1-XVIV.

and rapid change.[2] Man's ever expanding Faustian power over nature is, within this context, axiomatically assumed.[3]

Future-oriented planning is usually explicitly or implicitly based upon similar extrapolations from the present. It commonly involves an approach of "disjointed incrementalism," moving cautiously but deliberately in the direction of the presumably inexorable trend, modified only by the planner's individual, professional, or institutional conception of human betterment.[4] Such planning leads to the further elaboration of valued institutions, the gradual shifting of priorities, improvements in cost effectiveness, or the redesign of systems of service delivery. Planning rarely challenges the fundamental assumptions of economic and technological growth—usually identified with progress—upon which it is based[5]

If our futuristic assertions are correct, if change of historic proportions beyond the control of planners or politicians is nearly upon us, then planners cannot afford to continue to be preoccupied with efforts which amend and sustain the status quo. Neither disjointed incrementalism, nor the seemingly unending discussions of the power, role, and influence of planners, remain central issues. Among our major new tasks emerges the need to invent, and if possible to experiment with alternative social institutions and physical arrangements which are expressive of the actual or anticipated changed circumstances of life.

This is no essay in ecology and therefore it is not the place to enter into a detailed examination of evidence to the effect that there is no simple counter-technology to the current technology of waste. The answer is not bigger and better pollution control devices, but only less pollution. All conversion of fossil fuels to energy is a source of pollution of the environment, and nearly all methods of modern technological production (including anti-pollution devices) depend upon such conversion of energy. Even if "clean" atomic breeder reactors are made

---

[2] Herman Kahn and Anthony J. Wiener, *The Year 2000* (New York: McMillan Co., 1967) p. 6f.

[3] *Ibid.,* p. 409.

[4] The phrase disjointed incrementalism is from Charles E. Lindblom, *A Stategy of Decision* (New York: Free Press, 1963). It has been used in connection with social planning by Alfred Kahn, *Theory and Practice of Social Planning* (New York: Russell Sage, 1969), p. 337.

[5] As an example of this type of thinking on the national level see, *Toward Balanced Growth: Quantity with Quality,* Report of the National Goals Research Staff, Washington, GPO, July 4, 1970, also *Agenda for a Nation,* Kermit Gordon, ed. (Washington: Brookings Institution, 1968).

operative by 1980, as the AEC anticipates, the excess heat generated on the earth's surface as a consequence of increased energy utilization will remain a dangerous source of environmental contamination. The United States consumes more than one third of mankind's energy production. Its rate of energy consumption doubled in the past twenty years. It is unimaginable that there is no limit to this exponential rate of growth, and totally unthinkable that other, less developed nations might aim to achieve a similar level of technological development.[6]

In the Age of Retrenchment Western civilization will experience profound changes not only in the nature and scope of its technology, but also in its social and political organization. Imagine a work week of less than 15 hours, and a work life of only 20 years. Perhaps young people will be kept out of the labor force until the age of 35. If a zero rate of population growth is reached, and let us say 80 is the anticipated life expectancy of man, then the median age will be approximately 40. Politically, this is likely to be a very low change oriented society.[7]

The Director of the Oak Ridge National Laboratories has called for a "commitment to a permanent social order" so that nuclear safety can be assured.[8] Are there any among us who dare consider this an acceptable price to pay in exchange for continued technological progress? Even if there were no other restrictions, if there were no limits to growth, can we ever permit ourselves to settle for a social order in which the permanence of social institutions is a requirement for survival?

This question is raised because it highlights one of the most important issues which must come to the forefront in discussions of the future of planners and of planning: the obligation of planners to join, and often to take sides in the arbitration of moral issues.

"Innovate in order to simplify . . . otherwise, innovate as sparingly as possible" is the maxim proposed by Paul Goodman for the era that lies before us.[9] This does not mean the end of technology, but it does mean

---

[6] "Energy Challenges of the Future", *Science*, 177, 8 September, 1972, pp. 875-6.
[7] Robert Theobald has developed some of these themes in greater detail, vide, Robert Theobald (ed), *Social Policies for America in the 70's* (New York: Doubleday, 1968), pp. 149-170. Also, see the dramatic warnings contained in the Club of Rome's report, "Project on the Predicament of Mankind". Vide, Donella H. Meadows, *et al, The Limits to Growth* (New York: Universe Books, 1972).
[8] Alvin M. Weinberg, "Social Institutions and Nuclear Energy", *Science* 177, 7 July, 1972, pp. 27-34.
[9] Paul Goodman, *New Reformation; Notes of a Neolithic Conservative* (New York: Random House, 1970) p. 9.

the end of progress as we have understood the term for the past nearly
two hundred years. It means the refinement of technology and the
redefinition of its function in society. Within such a context, suggests
Goodman, technologists (and planners among them) will be called upon
to serve not only as experts, but also as moral philosophers.

There is, of course, no limit to the speculation that is possible
concerning the particulars of a Retrenchment Society. Assuming a
relatively absolute, environmentally imposed limit upon growth, what
are the likely consequences? There are, broadly speaking, two alternatives: to reduce (not stabilize) world population and continue technological expansion; or to plan for the rational, just, and equitable distribution
of the limited energy and material resources. It seems to be a choice
between barbarism and planning. Barbarism will attempt the elimination
of surplus people, such as Blacks and Orientals; planning means, first of
all, making decisions concerning the elimination of wasteful production.
There is another, politically more poignant way to make this point.
Problems of poverty and the maldistribution of the world's resources will
not be resolved until excessive wealth, too, is considered to be a major
social problem.

In preparation for a more planned society in which production is
controlled, and limited available resources are more justly allocated
among men, economists will have to invent a new system of economics,
an economics of non-growth.[10] It is likely that in such a society the
distribution of essential goods and services will not be allocated to the
market system as we know it.

Rexford Tugwell, in his draft for a revised constitution for the United
States, suggests the establishment of a Planning Branch at the highest
level of the central government.[11] But will such a reorganization of our
governmental structure with its high regard for planning serve as the
basis for new freedom and rediscovered democracy, as Tugwell intends,
or will it provide legitimation for greater authoritarianism in the name of

[10] Kenneth E. Boulding, "Economics of the Spaceship Earth", in *The Environmental Handbook,* Garrett di Bell, ed. (New York: Ballantine, 1970) pp. 96-101. Also, see the debate among French socialists concerning the potential political-economic implications of a non-growth economy. "Ecology and Revolution—A Symposium", *Liberation* 17:6 Sept. 1972, pp. 3-12.
[11] Rexford G. Tugwell, "Constitution for A United Republics of America", *The Center Magazine,* 3:5, September 1970, pp. 24-49.

expertise and rationality? Planners, especially in their proposed new capacity as moral philosophers, need to ask themselves this question. At what point does planning unjustifiably restrict freedom? To what degree can such a point be identified with specificity? Or, somewhat differently, what should be the limits of planning?

Let it be clear, we are not here speaking of the functional limits that planners, serving as advocates of competing constituencies, place upon each other. Nor are we addressing either individual, or more broadly human constraints of time or intellect. What is at issue here is a central point of moral philosophy: in principle, what are the areas of human life which, assuming they could be rationally controlled through planning, *should not* be so controlled? That is what is meant by the limits of planning.

This study, this exercise in social theory, proposes that the limits of planning may be defined in terms of what are called, communal organizations. The family, the community, the nation, are examples of communal organizations. Such organizations are characterized by the fact that they are not oriented toward a specific, defining goal. They comprise those areas where generalized, non-specific, non-coercive, expressive human interaction is most likely to take place. Communal organizations are to be contrasted with formal organizations, which are the more usual object of planners' concerns. The elaboration of the concept of communal organization and an analysis of the interaction of communal organizations with each other is central to this study. Eventually, though not within the present context, a complete analysis of the interaction of communal organizations with formal organizations is indicated. For the present, this is a first attempt to delineate theoretically that area of human interaction—an area which clearly extends far beyond the personal—which, in our capacity as moral philosophers, we shall designate as the area beyond the limits of planning.

In its analysis of communal organizations, this study will focus on the community, i.e. local societies.[12] There are several reasons for this choice. Communal organizations tend to be vertically mutually inclusive, and community occupies a central position between family and nation, from which it is possible to view both the included social systems, as well as

[12] The concept of communal organization is derived from George A. Hillery, Jr., *Communal Organizations* (Chicago: University of Chicago, 1968). Hillery, too, focuses on the community in his analysis.

the social system within which the community itself is included. The view of the community as a complex (i.e., multi-level) social system is basic to the analysis which follows.[13]

Communities are the largest units of society which, in the United States, have been the subject of relatively comprehensive planning. We especially have in mind new-towns, quasi-institutions such as retirement communities, and utopian communities. The three case examples in this study are drawn from among such communities. In the light of the theoretical approach which is developed here, the examination of the consequences of planning in such communities has its special fascination because, unlike the planning of a social agency or a transportation system, planners are here presumably planning for the unplannable: a communal organization. The task of the planners, in this case, is not entirely unlike that of a marriage broker of old: he brings the couple together, perhaps he succeeds in making the *shidduch,* but there is no guarantee that love will follow.

Whereas this book seeks to address itself primarily to the concerns of planners and community sociologists, its larger aim is to reach many others. Its hopes also to reach many of those who are thoughtfully searching for, and working toward, a more peaceful world and a more just social order. Most specifically, it is addressed to those who identify with the movement for non-violent social change. While in recent years planners have become increasingly aware of the political meaning of their actions, "the movement," as it is affectionately called by those who identify with it, has become more sensitive to the fact that political action in the form of demonstrations and individual acts of resistance is not enough. Today, the formulation of, and experimentation with alternative institutions in anticipation of the New Society is equally at the core of the non-violent movement. Thus, both planners and social activists are beginning to show small signs of convergence in their pursuit of similar objectives.

---

[13] The term complex social system is taken from Odd Ramsoy, *Social Groups as System and Subsystem* (Glencoe: Free Press, 1962) p.11f. Ramsoy analyzes the relationships between inclusive and included systems. We are concerned with a single system, the community, functioning simultaneously as an included and an inclusive system, i.e., a three-level complex social system.

This study, while it makes no attempt to design the elements of an alternative society, sets as one of its primary objectives a preliminary theoretical delineation of those social-structural principles which, in the opinion of the author, should undergird both the alternative institutions and the alternative New Society. The communal organizations, it will be suggested, may be viewed as non-violent institutions. A non-violent social order is one in which the communal organizations have priority, are normatively and cognitively emphasized over the formal organizations.

The communal organizations, as has been said, define the area beyond the limits of planning. They are non-violent institutions, and they have priority in the desired, alternative New Society. Logically, this leads to a dilemma: it means that the New Society cannot be planned. This important issue will be raised again in the concluding chapter of this volume.

This study attempts to serve as a formal sociological analysis on the one hand, as well as a value laden, political-philosophical statement on the other. There are many who would consider such a marriage of motives illegitimate. The author is more inclined to believe that sociology which is not, at the minimum, selfconsciously cognizant of its political function is either naive or dishonest. Conversely, political action which remains uninformed of social theory risks being at its best ineffective, and at its worst counterproductive.

Let it be perfectly clear that the validity of the body of this study is not dependent upon the futuristic assumptions and the political biases stated above. They have been given expression in this introductory chapter primarily in order that the motivations and the value preferences of the writer be maximally explicit.

Chapter 2 will commence with a summary of the theory of communal organizations. It is based upon a detailed examination of sociological theory which appears separately, in Appendix A. Theoretically oriented readers may want to substitute the reading of Appendix A, for Chapter 2.

The definition of community which occupies most of the third chapter is closely related to the previous analysis of communal organizations. When the community is viewed as a complex, three level system, logically, eight combinations of communal and formal organizations emerge. This typology of communities, as well as what will be called anti-communities, is set forth in Chapter 4.

In order to serve as a test of the theory presented in the previous chapters, a specific problem of change from one community type to another is posed in Chapter 5. This is a pilot study, and the purpose is not one of reaching quantitatively validated conclusions. The primary aim is to test the applicability and the usefulness of the concepts and methodology.

The community studies which follow in Chapters 6, 7, and 8 focus on three "deviant communities." Each community is first described, and then analysed separately in order to determine to what degree it conforms to the anticipated characteristics of its type. Levittown, Moosehaven, and the Bruderhof, each in a different way, represent attempts to create unique, separate, and ideal social systems within the framework of the present society.

General findings and conclusions, to the extent that they are possible on the basis of such a small sample, are presented in Chapter 9. The implications of this study for community theory, and for planning theory and practice, are discussed in Chapters 10 and 11. There we will return to some of the important questions raised in this introduction.

Finally, in the concluding chapter, we permit ourselves to speculate on the systematic nature of an alternative society. Contrary to the visions of some, this society can not be a perfect, utopian society in which only communal organizations thrive. Such a society, were it possible, would hardly survive a day. Rather, it is a society which is structurally much like our own, one which permits of a variety of community types. The difference derives primarily from the industrialization of alternative values. In the alternative society, communal organizations are more highly valued than formal organizations. In the alternative society, being is valued over doing, content over achievement, love over competition, and process over goal.

CHAPTER 2

# COMMUNAL ORGANIZATIONS—A SUMMARY

The theory of communal organizations serves as the basis for our systematic analysis of communities. This chapter is a summary of that theory, based on the original work of George A. Hillery and others, and developed in greater detail by the author. A more detailed discussion of the theory of communal organizations appears in Appendix A, at the end of this volume.

It must be confessed at the outset that the theory of communal organizations requires much additional refinement and empirical validation. This theory, like any theory, is but a conceptual tool; it is useful to the extent that it gives accurate descriptive expression to empirical reality and casts new insights upon it. Whereas in this chapter no attempt is made to verify the theory, the remainder of this study is to be viewed as precisely such a first effort.

During the past two decades several significant contributions have been made to augment and/or modify social systems theory as it was first comprehensively stated by Talcott Parsons and his students.[1] The theory of communal organizations, which is importantly influenced by several of these analyses, proceeds upon a similar path. As with any scientific theory, its validity is importantly related to the degree to which it is accepted, criticized, modified, and expanded upon by colleagues.

---

[1] For a valuable summary of varieties of approaches to social systems theory, though not including communal organizations, see, Robert R. Mayer, "Social Systems Models for Planners," *Journal of The American Institute of Planners* 38:3, May 1972.

### Social Systems

Social systems are the basic units of analysis in systematic sociology. The concept, as used here, may be applied to any patterned social interaction which persists over time. There are many kinds of social systems ranging from simple dyads, such as friendships, to large complex organizations, such as corporations. The theory of communal organizations proposes that most social systems may usefully be classified as either communal organizations, or as formal organizations (often called collectivities). Whereas no particular social system will in all respects satisfy all the criteria for one or the other type, it is hypothesized that most organizations fall into one category or another. One might imagine a continuum stretching from One modal type to another. Relatively few organizations are in the middle; most are concentrated near the ends.

We generally think of corporations, associations, national and local governments, etc., as formal organizations. Families, friendships, ethnic groups, communities, and nations are examples of communal organizations.

Our method of defining communal organizations will be to contrast them on a variety of dimensions with formal organizations. It is a task of identifying similarities and differences. The reader will be helped in following the discussion by keeping an eye on Plate 1.

### PLATE 1

#### SIMILARITIES AND DIFFERENCES BETWEEN FORMAL AND COMMUNAL ORGANIZATIONS
#### A SYNOPSIS

*SOME SIMILARITIES BETWEEN FORMAL AND COMMUNAL ORGANIZATIONS*

1. Both are solidary interactional social systems.

2. Both are relatively highly institutionalized in that they possess:
   a. developed normative structures;
   b. a high level of value consensus;
   c. patterned reciprocal role expectations.

3. Both may include subsystems of their own, as well as of their opposite type.

4. Sentimental collectivity orientation (loyalty, commitment) is a variable.

## DIFFERENCES BETWEEN FORMAL AND COMMUNAL ORGANIZATIONS

(Each of the dimensions may be considered as representing a continuum)

| *FORMAL ORGANIZATIONS* | *COMMUNAL ORGANIZATIONS* |
|---|---|
| 1. Oriented toward a specific, defining goal | 1. Not oriented toward a specific, defining goal |
| 2. Functional collectivity orientation | 2. No functional collectivity orientation |
| 3. Linked by contract, i.e., by specified and limited cooperation | 3. Linked by generalized cooperation (active and passive) |
| 4. Mechanistic Interaction | 4. Structured free wheeling |
| 5. A variety of roles and a formal hierarchy | 5. A variety of roles but no formal hierarchy |
| 6. Normative, utilitarian and coercive forms of power are legitimate | 6. Only normative power is legitimate |
| 7. Created externally or by its elements | 7. Generated by its elements |
| 8. The inclusive system defines the roles of the subsystems | 8. The inclusive system is defined by the subsystems |

### Some Similarities Between Formal and Communal Organizations

Both formal and communal organizations are interactional social systems which are "solidary," in the sense that they are not conceptually divisible. An example of the divisible type of system would be a market in which there is a buyer and a seller, or football game in which there are two teams, one in competition with the other. The analysis of such non-solidary social systems, as well as of coalitions and federations, is excluded from this discussion of communal and formal organizations.[2]

Both formal and communal organizations are better understood when viewed on at least two, and sometimes on three levels. Every social system consists of elements which interact. In micro-systems these

---

[2] Odd Ramsoy *op. cit.,* was the first to suggest the idea of markets and non-solidary social systems. Roland Warren has been concerned with the analysis of federations and coalitions. Vide, Roland L. Warren, "The Interorganizational Field as a Focus of Investigation," *Administrative Science Quarterly* 12:396-419, December 1967.

elements are individuals (or more specifically, individuals in roles). In macro-systems they are other social systems, i.e., subsystems. In theory, every role and every social system can be thought of as being vertically included within another social system, to the limits of the social universe. Thus any given system may be viewed as both an inclusive system and an included system. It is much like the layers of an onion, they both include and are included by others.

The linkages among the elements of a social system as, for example, between two members of a single organization, are thought of as *horizontal linkages.* The linkage between an element and the inclusive system is called a *vertical linkage.* This, for example, is the relationship between the member and the organization, *qua* organization. The linkage between an element of a system and another system or role which is beyond the boundaries of the inclusive system is called an external linkage. This is a case of the relationship between the member of an organization and a non-member, or between a member and another organization. (See Plate 11 in Appendix A).

Both communal and formal organizations, when viewed as inclusive systems, may include subsystems of their own type, as well as of their opposite type. For example, a business corporation includes both formal and informal subsystems. A community includes both families and business organizations. The communal elements and the formal elements both cooperate and are in contest with each other, leading to intrasystemic strain.

Both formal and communal organizations tend to be highly institutionalized. This means that reciprocal role expectations among their elements are highly patterned. They tend to possess well developed normative structures, i.e., right, traditional ways of action, and therefore, in system related matters, a high degree of agreement and of value consensus is to be expected. Families and bureaucracies are examples of highly institutionalized systems. They are to be contrasted with crowds or audiences, which possess relatively little institutionalization.

### The Differences Between Formal and Communal Organizations

As indicated above, the differences between formal and communal organizations are not to be viewed as absolutes. Though expressed as contrasting poles, they are differences only of degree. The eight analytical dimensions which are discussed below are, of course, related to each other. Perhaps the list could be condensed. It remains one of the central tasks in the further development and testing of this theory to determine to what degree each of these eight dimensions consistently supports the specification of a dichotomy.

1. Formal organizations are oriented toward a specific goal. By specific goal is meant an output, something that can serve as a measurable input into another system. In raising the question of orientation toward a specific goal, the focus in the analysis should be on the *orientation* of the system, rather than on the *goal,* the material or symbolic output. The idea of specific goal means to exclude short spurts of action directed toward a specific end, e.g., eating a meal to satisfy hunger. It also excludes generalized goals, such as happiness, love and freedom.

Communal organizations differ in that they are *not* oriented toward a specific goal. It is, for example, a perversion of the idea of family to say that it is oriented toward the specific goal of producing and socializing children. Many families have no children, others have completed the task of child rearing. Members of a family, if they were to give expression to goals, might speak of joy, or love, but not of outputs.

In order to avoid the pitfalls of absolutism, it is better to speak of formal organizations being high specific goal oriented, and communal organizations low specific goal oriented. For purposes of brevity, in the text which follows, frequent use will be made of the terms high goal oriented for formal organizations, and low goal oriented for communal organizations.

2. In formal organizations the elements function primarily in such a way as to contribute directly or indirectly to the specific goal of the system. This is called functional collectivity orientation. In communal organizations there is little functional collectivity orientation and the elements are not primarily to be understood in functional terms.

In formal organizations, as a bare minimum, the elements must function so as not to impede the high goal orientation of the system. For example, all of the formal subsidiaries of a bureaucratic organization are expected to make their contribution to the goal. Every department of a hospital is expected to contribute to the provision of health services. The "informal system" within the hospital, e.g., friendships between staff and patients, is encouraged or tolerated only to the degree that it either contributes to the goal, or remains functionally neutral.

Whereas formal organizations are integrated primarily by means of functional collectivity orientation, the integrative basis of communal organizations is derived primarily from another source, from shared symbolic identity. A proper name, shared traditions, common stigmas, or ritually valued possessions provide the basis for the integration of communal organizations. It is primarily the shared cultural heritage of rural Italy that holds together Americans of Italian descent. Symbolic ties of blood bind a clan. An exalted name, a flag, and shared heroic memories are the kind of images that bind a nation.[3]

There is another type of collectivity orientation which is often, but not necessarily, associated with the first. This second type is called sentimental collectivity orientation. It is the sentiment of loyalty, love, commitment, sense of duty, etc., which the elements express toward the inclusive system. Sentimental collectivity orientation is a variable which may apply either to formal or to communal organizations; it does not constitute a point of difference between the two types. It should be noted that this is not the same as saying that sentimental collectivity orientation is not an important variable. The point is simply that sentimental collectivity orientation is not a variable on the basis of which formal and communal organizations may be distinguished from each other.

Symbols of identity are to be found in both formal and communal organizations. They frequently serve as media for the generation or expression of sentimental collectivity orientation. In communal organizations, when sentimental collectivity orientation is low, the symbols of identity serve to integrate the system: in formal organizations the symbols of identity supplement the integrating function of specific goal orientation.

3. Formal organizations are linked both horizontally and vertically primarily by contract. A contract is an agreement to limit cooperation. By contrast, communal organizations are linked primarily by generalized cooperation. Generalized cooperation does not mean unlimited cooperation, but rather unspecified and undefined cooperation.

Generalized cooperation horizontally, among the communal elements of a social system, may take either an active or a passive form. When it is active, it is mutual aid, neighborliness, friendship, etc. When it is passive, it constitutes a relationship of "live and let live," of *laissez-faire.* Horizontal linkages between mixed types, that is between formal and communal organizations, may be either by contract or by generalized

---

[3] Nations are, of course, to be distinguished from states. Nations which fall short with respect to the establishment of shared symbolic identities have difficulty in sustaining unified governmental structures. Witness the struggles within some of the new states in Africa which have been somewhat insecurely superimposed upon the traditional tribal organizations.

cooperation. For example, a factory located in a community may have both contractual and neighborly links with families who reside in the vicinity.

Communal organizations, especially when they are viewed from the inclusive level, are vertically linked primarily by passive generalized cooperation. This, for example, is the nature of the linkage between a community and its subsystems. The multiplicity of types of horizontal, vertical, and external linkages which are possible within this kind of system becomes a dismayingly complex topic. The reader is again referred to Appendix A, for a more detailed discussion.

4. The nature of social action in formal organizations, ideally speaking, is mechanistic: for every action there is an expected, predictable reaction. Exchange theory, as developed by Blau and others, is applicable.[4] Such systems lend themselves readily to the calculation of presumed causes and effects, of give and take, and of conventional economic exchanges.

In communal organizations, by contrast, linkages of "structural free wheeling" predominate. In communal organizations the elements are linked to each other freely in such a way that a change in any one of them does not necessarily produce a change in any other. The ties are symbolic, rather than mechanical. Thus, I help my friend because I want to, and because he is my friend, not because I expect or want a favor in return. However, the linkage is not random: it is structured, since not every person in the world is my friend. Moreover, over time, patterned though not contractually specified ways of our expressing our friendship have evolved.[5]

5. The elements of social systems differ from each other in that they perform a variety of roles. In formal organizations the roles tend to be hierarchically arranged with the leaders, those exercising the most power, at the top. Statuses are usually clearly specified within such systems. In communal organizations, whereas they too may include a variety of roles, there are no formal hierarchies.

In an extended family, for example, there may be a mother and a father, daughters, grandmother, and uncles. Each person has his or her defined role, but ideally, all are equally members of the family. Since only idealized, "liberated" families function in such an egalitarian fashion, it might be more appropriate to cite the example of a commu-

---

[4] Peter M. Blau, *Exchange Power in Social Life* (New York: John Wiley, 1964).
[5] The idea of structural free wheeling may suggest the basis of a sociology of love, or differently, a sociology of creativity.

nity as a case in point. The sub ystems of communities are families, corporations, voluntary associations, governments, and many other social systems. The more hierarchical the elements of a community become, as in a dictatorship, the less the system resembles a communal organization. Similarly, in a family, the more hierarchical the structure, the more it resembles a formal organization.

Associated with the question of hierarchy and roles is the issue of leadership. Formal organizations have official leaders, mayors, presidents, chairmen. In communal organizations, to the extent that leadership is identifiable, it is informal, personal, and sometimes charismatic.

6. In formal organizations all three types of power identified by Etzioni, coercive power, utilitarian power, and normative power, are exercised with varying degrees of legitimacy, depending upon the actor and the situation.[6] By contrast, in communal organizations, only normative power is legitimate.

Power is the ability to exercise control over the actions of others. In social systems power is required for two purposes: in order to regulate the internal organization of the system, and in order to influence its environment. Coercive power usually implies the use or threat of physical force. Utilitarian power involves the exchange of material assets. Normative power is expressed primarily in symbolic terms. When exercised horizontally, it functions as what is usually called social control. When exercised vertically, it takes the more active form of persuasion.

Formal organizations may potentially use all three types of power for both internal and external purposes. A government trying to recruit an army is an example of a formal organization exercising all three types of power, internally. Attempts to recruit a voluntary army are conducted primarily by normative means, by calls to duty and to patriotism. However, soldiers are paid salaries, because it is assumed that patriotism alone cannot be relied upon. Finally, an AWOL soldier may be incarcerated. In its foreign (external) policy, a government may first attempt diplomacy, then economic sanctions, and if that also fails, military force.

Communal organizations, by contrast, have only the single option of calling upon normative power. When two lovers resort to coercion, then love is threatened. When the relationships among the subsystems of a community become enmeshed in an endless morass of laws, economic

---

[6] Amitai Etzioni, *The Active Society* (New York: Free Press, 1968) pp. 357-9.

exchanges, and reciprocal obligations, then the communal nature of the system is in danger of being undermined. In such a community "nobody does nothing for nothing."

In communal organizations power is more evenly distributed among the subsystems than in formal organizations—which is another way of saying that there is no hierarchy. Informal leaders have influence in communal organizations, and social control is exercised by peers.

7 & 8. In formal organizations the inclusive system defines the roles and functions of the subsystems. The leader, acting in the name of the inclusive system, allocates the tasks and duties. In communal organizations, by contrast, the inclusive system is defined by the subsystems. An ethnic group, for example, is only as vital as its members desire it to be.

It is common for a new formal organization to be created externally by the inclusive system of which it is to become a subsystem. This is the case of the branch store or of the new department. Under the circumstances, a new formal organization, such as a new social service agency or a small business organization may be generated by its elements. For example, several individuals may come together in order to charter a new association.

Communal organizations, by contrast, cannot be created by external systems, they must be generated by their elements. A family is created by a man and a woman, not by a mating agency. A community is generated by the people who live in a place; it is not created by city planners. It is true that the cultural form for a family, or for a community, pre-exist each specific instance of the generation of a new family or a new community. But the important point is that the elements generate a new system which previously did not exist in its specificity. By contrast, in formal organizations the system, in its specificity, may be designed externally. Not only in its form, but also in its concreteness, the new organization may be created largely without the action of its elements.

## A Definition of Communal Organization

Before concluding this chapter let us attempt the difficult task of proposing a definition of communal organization:

A communal organization is a relatively highly institionalized social system, differing from formal organizations in that it is characterized by low goal orientation, is internally linked primarily by means of generalized cooperation, and is normatively controlled by peers and by informal leaders.

# CHAPTER 3

# A DEFINITION OF COMMUNITY

Our purpose in this chapter is to suggest a definition of the word community which is maximally inclusive—which excludes as few as possible of the social units that have traditionally been called communities. It is important to emphasize that the word community is here used as the name for a specific type of human group. Unfortunately, this same word has also been used, at times, to refer to a sentiment, i.e., to the "sense of community." Community as sentiment is not under discussion in this chapter (see Plate 2).

## PLATE 2

### TWO MEANINGS OF THE WORD COMMUNITY

| Community as a human group | Community as sentiment |
|---|---|
| town | sense of community |
| city | commitment |
| folk village | loyalty |
| vill | communion |

Definitions are generally made with a view to their utility with respect to the task at hand. For example, Warren's definition of community, "that combination of social units and systems which perform the major functions having locality relevance," places the emphasis on the subsystems of the community, rather than upon the community.[1] Warren is primarily interested in an analysis of the horizontal linkages among formal subsystems, and the linkages of these subsystems with systems external to the community.

---

[1] Warren, *The Community in America* (Chicago: Rand McNally, 1963) p. 9.

**PLATE 3**

*THE THREE LEVELS OF COMMUNITY - A SCHEMATIC DIAGRAM*

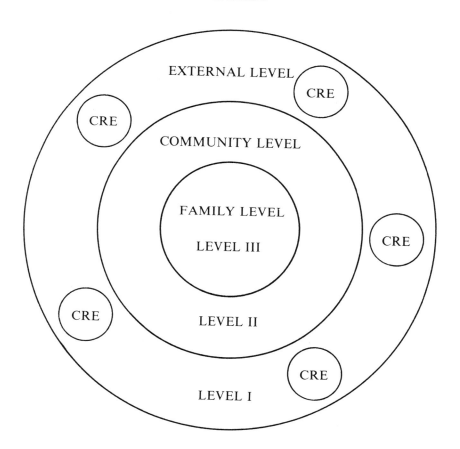

CRE = COMMUNITY RELEVANT ELEMENT

Our view proceeds from a perception of the community as a three-level, complex social system. The middle is the *Community Level*, the inclusive level is the *External Level*, and the included level will be called the *Family Level*. Conceptually, the three levels differ from each other primarily in terms of their degree of inclusiveness. (See Plate 3)

There are, of course, many definitions of community, each definition fulfilling its purpose within the context of an analytical system. There is no unanimity, but all or nearly all defintions of community include two main elements: a territorial concept and a concept of social interaction.[2] For our purposes, we find it best to develop a new definition:

*A community is a local society, a communal organization including formal and communal subsystems.*

### The Community as Local Society

The word local means to exclude, among other things, the nongeographic community, e.g., the scientific community. It refers to a contiguous physical area. Moreoever, it implies the existence of a larger unit of which the community is a subsystem. The larger unit can be thought of as the nation, or national society. The nation is a communal organization which, like other communal organizations, includes both formal and communal organizations as subsystems. It includes, e.g., corporations, associations, political parties, and governmental organizations, as well as communities.

Our interest is primarily in those elements of the nation through which it is linked with the community: the community relevant elements.[3] For purposes of community analysis, it is not of primary importance whether the nation as a whole behaves in all respects like a communal organization. More important is whether the vertical linkages between the community relevant elements and the community are expressive of an inclusive level of communal organization. With respect to each individual community, the community relevant elements are likely to differ.

[2] A long list of definitions of the word community has been collected by Ernest B. Harper and Arthur Dunham, eds., *Community Organization in Action* (New York: Association Press, 1959) pp. 23-27. Also see, George A. Hillery, "Definitions of Community: Areas of Agreement," *Rural Sociology*, Vol. 20, 1955, pp. 111-123.
[3] The community relevant elements of the nation are not to be confused with Warren's locality relevant functions.

From the point of view of the community, these elements are what has previously been called the external level.

The word society is introduced into the definition of community to indicate a social system with a relative degree of permanence, which includes a multiplicity of social institutions to serve basic human needs, such as production, reproduction, consumption, recreation, and education. It includes many, though not necessarily all, of the organizational forms essential for survival. All the summer guests of an island resort hotel may, under circumstances, be considered a communal organization. They are not a community, however, because the organization lacks permanence and it is relatively restricted in its provision for human needs. For similar reasons, the congregation of a church would not fall within our definition of community, despite the fact that it is relatively permanent. On the other hand, the Satmarer Chassidim of Brooklyn are a community because of the multiplicity of institutional bonds that tie them.[4] The Chassidim, as well as the Bruderhof which will be discussed in detail later, have not been tied geographically to a particular locality. But both Chassidim and Bruderhof retain most of the other characteristics of a community and are therefore included within the classification. In short, the line between community and other types of communal organizations is not precise. Marginal cases might be classified as one or the other.

### The Community as a Type of Communal Organization

The community, which is a communal organization, includes subsystems such as families, ethnic groups, and neighborhoods which are also communal organizations. This does not mean to say that all these units are by definition communal organizations. Some such units are insufficiently institutionalized to justify being called organizations, as for example some neighborhoods. Others, as for example certain ethnic groups, may become highly goal oriented and thus more nearly resemble formal organizations.

The most universal communal organizational subsystems of communities are families. But in modern societies, individuals, isolated from families yet constituting complete functional units, are becoming increasingly prevalent. Individuals are, in this sense, communal subsystems

---

[4] Israel Rubin, "Chassidic Community Behavior," *Anthropological Quarterly,* 37, 3, July, 1964, pp. 138-148.

of the community.[5] Moreover, other types of communal subsystems, besides the family and individuals, are today of increasing importance, such as ethnic groups, occupational groups and age groups. For purposes of this analysis, however, the family will generally serve as the primary and representative subsystem of the community.

Families are linked to each other, i.e., linked horizontally, primarily by means of generalized cooperation. The horizontal links of families to formal organizations may be either by means of generalized cooperation or by contract. However, in Western society, the contractual links are usually dominant.

Families are vertically linked to the inclusive system, i.e., to the community, by means of passive generalized cooperation. There is no contract between the community and its families. The community cannot exercise coercive or utilitarian power over its families as long as it remains a communal organization. However, in the event that it does attempt to exercise such power, it begins to move in the direction of one type of deviance. This point will be explicated more systematically below.

Among the formal organizational subsystems of communities are business corporations, political organizations, social welfare organizations, voluntary associations, etc. These subsystems of the community are usually linked with each other and to extracommunity systems by means of contract. However their vertical linkage to the community is by passive generalized cooperation. Horizontally, they are linked with communal organizations both by means of contract and by passive and active generalized cooperation.

Leadership on the community level, as long as it remains low goal oriented, is informal and non-hierarchical. Theoretically, *from the point of view of the community,* whereas some of its subsystems are more inclusive than others, all of them (formal as well as communal subsystems) are equal—not in status or role, but in their rights.

Under some circumstances, a community may become highly goal oriented and may lose most of its characteristics as a communal organization. For example, the community may appear to merge with its parallel formal organization, the local government. But in such an event the community does not cease to exist; it merely assumes one of the

---

[5] Strictly speaking, it is of course incorrect to refer to individuals as social systems.

forms of deviance. At the other extreme, instead of becoming highly goal oriented, a community may devolve into a system with such low level of institutionalization *on the community level* that it appears to have ceased to exist. This is a different form of deviance.

Deviant communities shall be discussed in greater detail in the next chapter. By definition they fall within the general classification of communities, not outside it. They are called deviant because they fail to exhibit the characteristics of low goal orientation on all three levels. Whereas deviance might also arise from the fact that a community is not locality based, like the Chassidim, or is not sufficiently permanent or multifaceted, these other modes of deviance are not of primary concern in this study.

### *The Community Viewed as A Complex, Three-Level Social System*

Up to this point, we have examined the community primarily as a simple, one-level system without putting into equal focus either the External Level or the Family Level. A more adequate view will be gained by means of the three-level complex system approach. For purposes of simplicity and clarity we shall designate the three levels as follows:

Level I     - External Level
Level II    - Community Level
Level III   - Family Level

Level II is included in level I, and level III is included in level II. Hypothetically, each of the three levels may be either a communal organization or a formal organization, i.e., it may be either high goal oriented or low goal oriented. This leads to the potentiality of establishing a logical maximum of eight combinations of high and low goal orientation on three levels. Each of these combinations will be discussed in the next chapter.

The above definition of community establishes a theoretical norm, such that level I is low goal oriented, level II is similarly low goal oriented, and level III includes non-coercively, non-utilitarian controlled, low goal oriented subsystems, i.e., families. It assumes that the levels are vertically linked by means of passive generalized cooperation and are normatively controlled by informal leaders and peers. Formal organizations are included at each level, but they are not central to this analysis of communities.

## A Methodological Note

In this discussion we will consistently identify level I with the external systems, level II with the community, and level III with the subsystems of the community. But the application of the three-level systems approach which is here suggested is not necessarily limited to this model. For example, in a slightly different application, level I would be the community, level II the neighborhood, and level III the families. Another application might be, level I as the international "community" of all mankind, level II, the nations, and level III, local communities.

But the focus of our discussion, for the reasons given in the Introduction, is the community, i.e., local societies as defined in this chapter. Therefore the referents of the three levels are as designated above. The three-level systems approach and the high-low goal oriented variable, when applied to any of the alternate referents would, of course, lead to other interesting studies of social organization.

# CHAPTER 4

# A CLASSIFICATION OF COMMUNITY TYPES

The classification of eight community types which follows is deductively derived from the theory of communal organizations and the definition of community. They are all to be understood as "ideal types," in the tradition of Max Weber.[1] Like Platonic ideals, ideal types do not exist in reality; they are prototypes against which empirical reality may be measured. The examples that are mentioned in connection with each type are only illustrations; they have not been tested for fit.

In following the discussion in this chapter, the reader will find it helpful to keep an eye on Plate 4. We will first discuss the three *Deviant Communities:* the *Administered Community,* the *Intentional Community,* and the *Designed Community.* These will be contrasted with the *Crescive Community* on the one hand, and with the four types of Anti-community on the other: *The Total Community; Orwell, 1984;* the *Solipisistic Community;* and *Totalitarianism.*

### Deviant Communities[2]

In Deviant Communities two levels are high goal oriented and the third level is low goal oriented. The two high goal oriented levels are linked to each other by means of a *partnership.*

A parnership is a contract between two high goal oriented levels of a community. There are two types of partnership: solidary and reciprocal. In a solidary partnership the two high goal oriented levels are contracturally linked into one system, oriented toward a single, specific goal. In a reciprocal partnership the two high goal oriented levels are contractural-

---

[1] Hans Gerth and C. Wright Mills, *From Max Weber* (New York: Galaxy, 1958) pp. 59-60.
[2] It is important to emphasize that the word deviant as used here and throughout this study is not meant in a pejorative sense. Deviant means, simply, differing from the norm.

ly linked into a market-type system. In a market-type system the major elements (in this case, the two levels) are oriented toward reciprocal goals, as in a commercial exchange where one buys and the other sells.

In the *Administered Community* the external level and the community level, in solidary partnership, are high goal oriented, and the family level is low goal oriented. A church sponsored retirement community, a 19th century company town, a frontier stockade, an Israeli development town, are examples of administered communities. In these cases, level II, the community level, has a goal for which it has contracted with the external level, e.g., the social isolation and care of deviant individuals, the conquest and exploitation of new territory, the acculturation of new immigrants. The external and the community level relate to each other as inclusive and included formal organizations. Elements on level III, the families, suffer the fate of communal organizations which are included within a formal organization, they are limited and controlled by the inclusive system.[3]

The second type of deviant community is the *Intentional Community*. In these communities, the community and the family level, in solidary partnership, are highly goal oriented, and the external level is low goal oriented. Many 19th century utopian communities and contemporary autonomous religious societies might fit into this category. The goal in these cases usually has an ideological base, a specific social doctrine or a special concept of salvation.[4] Intentional Communities tend to deviate from the norms of national society. In their ideology, as well as their day-to-day life, such communities set themselves apart from level I. The external level, i.e., the nation's community relevant elements, is here low goal oriented. The best proof of this is the fact that the nation tolerates and does not overtly attempt to control these deviant communities. Thus, the relationship is one of passive generalized cooperation. When and if the national society becomes less tolerant, i.e., more goal oriented with respect to the Intentional Communities, then the community is easily destroyed. This, as we shall see, is what nearly happened to the Bruderhof when it was forced to flee Germany.

In the *Designed Community*, the third type of deviant community, the external and the family level, in partnership, are highly goal oriented and

---

[3] Hillery has proposed the following hypothesis which is in consonance with this analysis: "If a human group is primarily oriented to the attainment of a specific goal, then it excludes familial behavior." George A. Hillery, Jr., "Freedom and Social Organization: A Comparative Analysis," *American Sociological Review*, 36, (1971) :52

[4] The presence of an explicit ideology is not to be construed as a defining characteristic of Intentional Communities, but it is a probable and frequent, if not usual, concommitant.

the community level is low goal oriented. Private suburban housing developments, Levittown, and Columbia, Maryland, may be classified here. Whereas in the first two types of deviant communities, the partners have a solidary goal, in this case, the two levels have reciprocal goals, as in a market system. The external level in this case may be represented by a government agency, or by a real estate developer. The goal of redistribution of population may apply in the case of the government agency, and financial profit is the goal of the private developer. On the family level, the reciprocal individual and family goals dictate: e.g., a suburban house, a socially and culturally homogenous neighborhood, a high status environment, etc. Level II in the examples given is low goal oriented and is largely unrelated to the goals of levels I and III.

A government agency which effects population redistribution, it will be quickly agreed, is a high goal oriented community-relevant element of the nation. It is a bit more difficult to recognize that private contractors whose goals are financial profits, have a similar systemic role. There is no need to pursue here the essential relationship between developers and the national investment and banking system. The important point for our analysis is that the goals of the developers are viewed from levels II and III as an expression of the external system, and therefore, of level I. The community level is in this case low goal oriented, and suffers the fate of a communal organization included within a formal organization.

The idea of high goal orientation on level III requires additional comment. Is not the family inevitably a communal organization, rather than a high goal oriented formal system? This is a speculative point. Might it not be said, however, that when families are viewed primarily as production units for children, as among black slaves in colonial America, that they become high goal oriented? The families in rural southern Italy studied by Banfield may be a case in point.[5] For them survival has become the one goal in terms of which all actions must be judged. Similarly, some modern American families which are excessively production-consumption oriented may be approaching this theoretical ideal type, i.e., the non-communal family.[6] The stress between the

---

[5] Edward C. Banfield, *The Moral Basis of a Backward Society* (New York: Free Press, 1958).
[6] The highly goal oriented family or individual is much like Etzioni's "inauthentic individual." Such persons are not to be confused with those who are alienated, persons who reject the norms and values of society. Rather, they are true believers, plagued by doubt. They are persons who battle windmills, half aware of the fact that that is exactly what they are doing. The alienated man feels manipulated, the inauthentic man is aware of the fact that he is participating in his own manipulation. *Vide,* Amitai Etzioni, 1968, *op. cit.,* pp. 633-634.

economic and social aspirations of middle class American families, and the demands of parenthood and conjugal love is too familiar to require elaboration here. Our point is that they may be viewed systematically, as an example of intra-systemic strain, and as an expression of the reciprocal contract between the family level and the external level.

Before leaving the discussion of Designed Communities, because of their topical importance, we may mention in passing that urban ghettos appear to fit into this same classification. For the moment, let us make a methodological shift, such as suggested at the end of Chapter 3. In the case of the ghetto, the landlords, City Hall, and the externally controlled social agencies, are viewed by the residents as level I. The residents of the neighborhood are level III. In their effort to survive, they seek to meet the expectations and demands of level I. The ghetto family's day-to-day struggle for survival constitutes a highly goal oriented activity. There are reciprocal goals between level I and III: level I exploits level III for profit and level III tries to survive. Level II has been low goal oriented in the ghetto until recent years. With the reassertion of level II, the question is, will it become high goal oriented, i.e., politicized *within* the national society, or will it move in some other direction. The Intentional Community discussed above is an alternative, haltingly being attempted by Floyd McKissick in Soul City, North Carolina.

### Crescive Communities

*Crescive communities* are not deviant communities; they differ from them in that they do not contain a partnership between two high goal oriented levels. In this category all three levels are low goal oriented. A Crescive Community is a communal organization, within a communal organization, containing communal organizations. Thus, the Crescive Community comes closest to meeting all the criteria of community discussed in Chapter 3. The traditional village, town, or city may serve as examples of this kind of community. The question, to what degree can various American communities be classified under this rubric, will not be addressed in this study.[7]

The detailed study of the Crescive Community within the context of the theoretical framework which has been here proposed is beyond our present scope. As has been indicated earlier, our major interest lies with

---

[7] This is a very important question. Unfortunately, the empirical analysis of Crescive Communities poses problems which are much more complex than the study of deviant communities. As in medicine, it is easier to define illness than health.

the deviant communities. But before we can return to these, we must briefly examine the four anti-communities.

### Anti-Communities

There are four types of anti-community. In three of the four types, *Orwell 1984,* the *Total Community,* and the *Solipsistic Community,* one of the three levels in high goal oriented, and the other two levels are low goal oriented. In the fourth type of anti-community, *Totalitarianism,* all three levels are high goal oriented. In the first three anti-communities there are no partnerships between levels. The fourth case differs in that it constitutes total cooperation, i.e., contracted and unlimited cooperation among all three levels. Contracted and unlimited cooperation implies the total absence of opportunity *not* to cooperate.

The absence of partnerships between levels among anti-communities highlights a curious phenomenon. The anti-communities, as shall quickly become evident from the discussion which follows, are functionally more deviant than the deviant communities. Yet, structurally, the deviant communities appear to be more deviant than the anti-communities. Whereas three of the four anti-communities are high goal oriented on only *one* level, the deviant communities all are high goal oriented on *two* levels.

The most plausible explanation for this phenomenon derives from the fact that, in the deviant communities, the two high goal oriented levels must reconcile many of their differences before they can be joined in partnership. As a consequence of this reconciliation, the effect of the high goal orientation of the partnership upon the third, low goal oriented level is somewhat muted. In the anti-communities by contrast, no such reconciliation is required and the single, high goal oriented level becomes exceedingly powerful and dominant.

When level I is high goal oriented and the other two levels are low goal oriented, then a social system results which finds its most accurate description in George Orwell's novel, *1984.* In this case the community and the family level are likely to have such a low profile that it might be best said that they are excluded. In this anti-community, local societies as well as local subsystems, i.e., families and friendship groups, disappear. Everything is controlled by the nation (i.e., the national government) in the service of its goal.

The *Total Community* results from high goal orientation on the community level and low goal orientation on the family and the external level. Again in this case, levels I and III may be said to have been excluded. This anti-community is much like a total institution that prohibits the development of all subsystems, and that has lost its functional basis within the national society. Its goal is no longer related with level I purposes, e.g., the control of deviant individuals. Rather, it establishes its own goals, its own world within itself. The French prison colonies off the coast of Latin America about which Papilion writes may be a case in point.[8] Aldous Huxley's *Brave New World* is another, but different, example of the Total Community.[9]

The anti-community in which the family level is high goal oriented to the exclusion of the community level and the external level we will call the *Solipsist Community*. In this case there is a total fragmentation of society. Only individual, and perhaps family goals have salience. In such a society cooperation among men is not possible, except perhaps within small groups, as within families. The external level and the community level are excluded because, in the absence of all forms of cooperation they cannot exist.

The fourth and last anti-community that is logically possible according to the model, is the case in which all three levels are high goal oriented. We have chosen to call this anti-community *Totalitarianism*. In this case there is total cooperation among all three levels. This is a society—if indeed the word applies at all—which is totally, as if mechanically, integrated from level I, to level II, to level III, in the pursuit of a consistent set of goals. All actions on all levels are directed toward goal attainment and there is no room for included communal organizations.

In reviewing this catalogue of anti-communities it becomes evident that these are not communities at all. They are maginal, hypothesized states of human existence. All except the Solipsistic Community postulate a strictly ordered, coercive society. For the individual, it probably makes little difference what systemic level exercises the coercive power that controls him. All four anti-communities have in common the fact that at the family level there is no possibility for horizontal linkage by generalized cooperation. This is another way of saying that in each case man has been substantially dehumanized.

---

[8] Henri Charriere, *Papilion* (New York: Morrow, 1970).
[9] Aldous L. Huxley, *Brave New World* (New York: Harper, 1946).

## Planned Versus Historical Communities

The model is now complete. In its summarization in Plate 4, the crescive and the deviant communities are located at the center, and the anti-communities at the corners. Administered and Designed Communities have in common the fact that level I is high goal oriented, i.e., the external level imposes its control upon the community level or the family level. This end is generally accomplished by the inclusion of non-residents, advocates of the external level, within the community. The reference here is to individuals variously called planners, developers, administrators, or, in short, staff. They perform a linking function between the local society and its subsystems, and the national society. They implement the contract between the two high goal oriented levels. In other words, *Planned Communities,* as these two deviant types of communities shall be called, are characterized by the fact that there is a staff-resident split.

The staff is a part of a high goal oriented, hierarchical external system exercising coercive, utilitarian and/or normative power. The residents of planned communities are usually in agreement or in compliance with the goals of the staff, in the case of the Administered Community because they have a low goal orientation, and in the case of the Designed Community because they are a part of the partnership and have reciprocal goals.

Intentional Communities and Crescive Communities have in common the fact that they both function within a national environment whose community relevant elements are low goal oriented. There is no staff-resident split. These two types will be called *Historical Communities.* Leadership at the community level, to the extent that it is identifiable in the Crescive Community, is likely to be informal or charismatic rather than formal and official. In Intentional Communities leadership is official, but local, i.e., not an advocate of the external level. In this sense, the Historical Communities are more democratic in their structure than the Planned Communities.

The difference between Administered Communities and Intentional Communities on the one hand, and Designed Communities and Crescive Communities on the other, is, as Plate 4 clearly indicates, that the first pair has high community level goals, and that the second pair has low community level goals.

**PLATE 4**

*A CLASSIFICATION OF COMMUNITY AND ANTI-COMMUNITY TYPES*

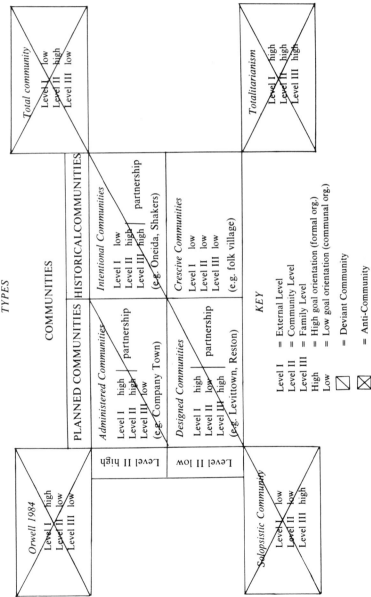

**COMMUNITIES**

*Total community*

| Level I | low |
| Level II | high |
| Level III | low |

*Totalitarianism*

| Level I | high |
| Level II | high |
| Level III | high |

**PLANNED COMMUNITIES** | **HISTORICAL COMMUNITIES**

*Administered Communities*

| Level I | high | |
| Level II | high | partnership |
| Level III | low | |

(e.g. Company Town)

*Intentional Communities*

| Level I | low | |
| Level II | high | partnership |
| Level III | high | |

(e.g. Oneida, Shakers)

*Designed Communities*

| Level I | high | |
| Level II | low | partnership |
| Level III | high | |

(e.g. Levittown, Reston)

*Crescive Communities*

| Level I | low |
| Level II | low |
| Level III | low |

(e.g. folk village)

Level II high

Level II low

*Orwell 1984*

| Level I | high |
| Level II | low |
| Level III | low |

*Solopsistic Community*

| Level I | low |
| Level II | low |
| Level III | high |

**KEY**

Level I    = External Level
Level II   = Community Level
Level III  = Family Level
High       = High goal orientation (formal org.)
Low        = Low goal orientation (communal org.)

◸  = Deviant Community

⊠  = Anti-Community

## An Afterthought

It may be useful to compare this model with Etzioni's.[10] The "mobilized" nation-state is the primary unit of analysis for Etzioni, whereas in our case it is the local society. This is an important distinction not only because the nation-state in our language is a formal organization, but also because the focus is on a much larger, more remote unit of action. Etzioni's concern with the "mobilized," i.e., the goal oriented potential of social systems, leads him to emphasize the control and consensus achieved by "guidance systems" which coordinate from above. Communal organizations, by contrast, emphasize the generalized cooperation that each social unit must discover within itself, concurrently weighing and frequently censoring the goal oriented behaviors that non-communal units attempt to impose upon them. The difference between Etzioni's approach and ours is a difference not only of concept, but also of values. It is somewhat like the oversimplified contrast between the philosophies of West and East: the emphasis upon doing and achieving, as against the emphasis upon being.

---

[10] Etzioni, *op. cit.*, pp. 466-467.

# CHAPTER 5

# THE MEASUREMENT OF CHANGE ON THE COMMUNITY LEVEL

The model of community types which was set forth in the previous chapter requires empirical validation. One way to accomplish this end would be to conduct detailed studies of a large number of varied communities, distributed among each of the eight community and anti-community types. But the approach which has been undertaken in this study is more modest. Only three empirical cases, each exemplifying one of the three deviant community types, have been examined in detail and tested for fit against the model. The focus is on the community level, with less emphasis upon the family level and the external level. The result is a pilot study which serves primarily as a preliminary test of the theory and the methodology.

In this chapter a hypothesis concerning chage on the community level will be introduced. Its purpose is only in part to test the specific question. Equally important is our desire to illustrate two points: 1) that the theory and the model may be usefully applied in empirical research, and 2) that the model, though it has been presented in a static manner up to this point, is to be applied dynamically, within the context of never ending social change.

While proceeding along this route, the larger objectives of this study are not forgotten. The aim is to identify and specify that area which is beyond the limits of planning. Each of the deviant communities has, in the broad sense of the word, been partially planned. These communities therefore represent instances of planning in the area of communal organizations which, hypothetically, is beyond the limits of planning. Therefore, in examining these three communities the question is, what has been the practical outcome of planning the presumably unplannable. And what has been the result of these planning efforts over time?

### A Hypothesis

Deviant communities may hypothetically be said to be unstable within their category. They are mixed types. Their partnerships, which are formed as a result of the contractual linkage between their two high goal oriented levels, are likely to deteriorate as a result of one of the two levels taking control, and excluding the other. In such a case, the deviant community becomes an anti-community. The other possibility is that the partnership between the two high goal oriented levels erodes and that the deviant community begins to resemble a Crescive Community.

It is hypothesized that: *Deviant communities tend to change over time, either in the direction of Crescive Communities, or in the direction of anti-communities.* Whether communities are more likely to move in one direction than the other is an important secondary question which is not pursued in this study.

Deviant communities, in order to move in the direction of the Crescive Communities, are expected to exhibit two processes which are called *goal erosion* and *democratization.* The combination of these processes is represented by a movement downward and toward the right in Plate 5. Goal erosion means that community level goals become less pronounced over time, and democratization means that official leadership representing the external level is eliminated, i.e., the elimination of the staff-resident split.

The process of *decommunization,* i.e., the movement in the direction of

## PLATE 5

### *AN OPERATIONAL CLASSIFICATION OF COMMUNITY TYPES*

| PLANNED COMMUNITIES | HISTORICAL COMMUNITIES |
|---|---|
| *ADMINISTERED COMMUNITIES* | *INTENTIONAL COMMUNITIES* |
| High Level II goal<br>Staff-resident split | High Level II goal<br>No staff-resident split |
| *DESIGNED COMMUNITIES* | *CRESCIVE COMMUNITIES* |
| Low Level II goal<br>Staff-resident split | Low Level II goal<br>No staff-resident split |

the anti-communities, is characterized by the increase in goal orientation on *one* of the three levels, to the exclusion of the other two levels. It is important to note that decommunization is not simply the direct opposite of goal erosion and democratization. In Plate 4, it is *not* simply a movement upward and to the left. Rather, it is represented by a movement toward the corners, in any one of the four directions, toward one of the anti-communities.

### Defining the Variables

*Goal Erosion*   Goal erosion refers to the degradation, dissolution, or overt rejection of previously institutionalized community level specific goals. Goal erosion can, of course, take place only within Administered or Intentional Communities because these are the only deviant community types which are high goal oriented on level II, the community level.[1]

If level II is characterized by high specific goal orientation, then we would expect to be able to identify a clearly, probably overtly expressed goal or set of goals for the community. The goal of level II is expected to be identical with the goal of the solidary partnership of which it is a part. It will be recalled that a solidary partnership results from a contract between two high goal oriented levels and that it functions in the manner of a single, high goal oriented system. In the Administered and the Intentional Community, in identifying and specifying level II goals, we therefore need to examine the goals of the partnership. Moreover, within the partnership system we would expect to be able to identify a formal hierarchy, the exercise of coercive, utilitarian, as well as normative power, the attempt to define and control communal organizational subsystems, and evidence of contractual vertical linkages between the two levels which participate in the partnership.

If level II is low goal oriented, as in Designed Communities, then we would not expect it to be a member of a partnership. It would lack a distinct, identifiable goal, leadership would tend to be informal and nonhierarchical, exercising primarily normative power, and the vertical linkage between the levels would be characterized by generalized cooperation and structural free wheeling. (See Plate 6)

---

[1] Goal erosion might also take place in anti-communities. But that is beyond the scope of this study.

## PLATE 6

### *THE DIMENSION OF LEVEL II GOAL ORIENTATION IN COMMUNITIES*

| *HIGH GOAL ORIENTATION* | *LOW GOAL ORIENTATION* |
|---|---|
| 1. Specific, identifiable level II goal | 1. No specific, identifiable level II goal |
| 2. Formal leadership hierarchy; formal leaders | 2. No leadership hierarchy; informal leaders |
| 3. Coercive, utilitarian and normative power | 3. Normative power only |
| 4. Vertical linkage by contract | 4. Vertical linkage by passive generalized cooperation; no vertical contracts |
| 5. Limited communal organizational subsystems (families) | 5. Unlimited communal organizational subsystems (families) |
| 6. Roles of residents defined by the community (level II) | 6. The community (level II) defined by the residents |

*Democratization*    Democratization has been previously defined to mean, within the present context, the elimination of the staff-resident split. In Plates 4 and 5, it is represented by movement from the left to the right, i.e., the Planned Communities become Historical Communities.

Staff are individuals with formal authority within the community who represent the interests, i.e., the goals, of level I. Such individuals may be planners, developers, officials or other agents overtly representative of the external system or systems, who may or may not reside within the community. In some cases they may be individuals who appear to be residents, who did not originally function as staff and who are not generally recognized as such, but who have been effectively coopted by an external system and therefore fill a staff role. The central criterion is not professional ability or status, but that the individuals identified as staff express and promote level I goals.

The residents are individuals or families who constitute level III. Whereas in Designed Communities the roles of resident and staff may be compatible within a single individual, in Administered Communities quite the opposite is to be expected. This is because in Designed Communities both staff and residents (level I and III) are high goal oriented, whereas in Administered Communities one is high and the other is low goal oriented.

In deviant communities which are characterized by a staff-resident split, i.e., in Planned Communities, we expect either solidarity or reciprocal complementarity between the goals of level I and the high goal oriented level II or III, i.e., within the partnership. In Administered Communities the partnership is solidary; in Designed Communities it is reciprocal. In both community types, staff, formal representatives of level I, exercise normative, utilitarian, as well as coercive power in the pursuit of external level goals within the partnership (see Plate 7).

In communities where there is no staff-resident split, we would not, of course, expect to find the inclusion of officials or of coopted representatives of level I. The vertical linkage of such communities to the external level would be one of passive generalized cooperation, the external systems exercising primarily normative power over the community.

It is to be noted that democratization as here defined refers to the vertical linkage of levels II or III, *to level I*. It does not refer to the vertical linkage between levels III and II. In other words, democratization is a measure of local society (community) or family external to the external system; it does not measure individual or family linkage to the community.

## PLATE 7

### *THE DIMENSION OF STAFF-RESIDENT SPLIT IN COMMUNITIES*

| *PRESENCE OF STAFF RESIDENT SPLIT* | *ABSENCE OF STAFF RESIDENT SPLIT* |
| --- | --- |
| 1. High level I goals | 1. Low level I goals |
| 2. Level I in a reciprocal or solidary partnership | 2. Level I not in a partnership |
| 3. Presence of staff representing level I | 3. Absence of staff representing level I |

| | |
|---|---|
| 4. Contractual vertical linkage to level I | 4. Vertical linkage to level I by passive generalized cooperation |
| 5. Roles of levels II or III defined by level I | 5. Level I defined by levels II or III |
| 6. Staff exercises normative, utilitarian, and coercive power | 6. Only normative power exercised by level I |

*Decommunization*    Decommunization takes place when a community moves in the direction of one of the anti-communities, i.e., when one of the three levels becomes highly goal oriented and dominant, to the exclusion of the other two levels, or, in the case of Totalitarianism, when all three levels become highly goal oriented.

All anti-communities have in common the fact that horizontal linkage among level III units is either characterized by contract, or is totally absent. In other words, active generalized cooperation among families and individuals is precluded.

## PLATE 8

### *THE DIMENSION OF DECOMMUNIZATION*

1. High goal orientation on only one, or on all three levels.
2. Absence of active generalized cooperation among families.

### *The Selection of Indicators*

The process of goal erosion, democratization and decommunization have, to this point, been discussed only in ideal terms. But in the examination of empirical cases it is to be expected that phenomena will fall short of the ideal types. Moreover, it will be remembered that the two types of organization, formal and communal organizations, are understood to differ in degree and that therefore each of the variables should be assumed to be continuous in nature.

In examining each of the cases for fit, the overall *pattern of conformity* to the model needs to be taken into consideration. It can not be expected that an empirical case will fully conform to the model in each of the variables. Occasional deviations are anticipated. However, when important deviations occur, then they need to be scrutinized in order to determine their origin, or to uncover fallacies within the theory or the model.

The two columns in plates 6 and 7 are assumed to be reciprocals of each other, which is another way of saying that the represent two ends of a single continuum. In practice, it is possible to determine, e.g., whether there is a formal leadership hierarchy, or whether there are vertical contracts. However, the absence of hierarchy and the presence of passive generalized cooperation can usually be measured only by indirection.

Each of the variables has been operationalized by means of multiple indicators. In the selection of these indicators emphasis has been placed throughout upon the identification of structural, rather than motivational or psychological characteristics. The indicators have been organized into common-sense categories and are to be found in the Community Schedule which appears at the end of this volume, in Appendix B.

The focus of this study is on level II; it is primarily a study of the community level. Family and societal variables are reflected only to the degree that they are thought to be of significance for the analysis of the community level. A more complete analysis of level III would require a survey of family structures within each community setting with the aim of determining whether families are predominantly high or low oriented. Such an undertaking was clearly not feasible within the context of the present study. Survey data on families were used only when available from other sources, but then their results were not always relevant to the issues at hand.

The fact that the family level was not fully, systematically analyzed is, of course, an important limitation of this study. On this level, the discussion of each case depends upon incomplete and inferential data.

A similar, but different point must be made with respect to the analysis of level I, which in a sense includes all of American society. It it totally beyond the scope of this study to attempt to determine whether and in what manner contemporary American society is high or low goal oriented. Some general, speculative inferences will be drawn in the concluding chapters, but these are not based upon the empirical portions of this study.

### Selection of the sample

Three communities were selected, one representing each type of deviant community. The cases were all drawn from the American scene since preliminary investigation has shown that many cultural varibles are likely to intervene in an international comparison of deviant communities.[2] In selecting the sample, Moosehaven, Bruderhof, and Levittown, New York, a number of criteria and limitations had to be taken into account:

1. Preference was given to older, over younger communities. A study of short-lived communites, and especially a study of the reasons for their demise, might have led to different conclusions.

2. Since each case was to be examined ove. time, only communities could be included for which extensive secondary sources of an historical nature were available.

3. Communities received their preliminary classification on the basis of their assumed structural characteristics at the present moment in time. Thus, the first task in each case was to verify this preliminary classification. Thereafter, this classification was contrasted with the historical data. A longitudinal study, which might have been preferable, though not essential to this analysis, was not possible.

4. Preference was given to communities which are currently accessible. Geography, travel costs, and availability of personal contacts were taken into consideration in making the final selection.

The cases that were originally considered for inclusion in the sample were as follows:

*Administered Communities*

> Pullman, Illinois
> Hershey, Pa.
> Moosehaven, Florida
> Lowell, Mass.

*Intentional Communities*

> Bruderhof
> Oneida
> Amana

*Designed Communities*

> Park Forest, Illinois
> Columbia, Maryland
> Greenbelt, Maryland
> Levittown, New York

---

[2] Shimon Gottschalk, "Citizen Participation in the Development of New Towns: A Cross National View," *Social Service Review,* 45;2, 1971.

### Data Collection

The method of data gathering differed somewhat in each case. First, secondary sources were carefully examined. They constitute the most important source of data in each case. Then, each community was visited by the investigator. In each case, additional, secondary sources became available on site.

In Moosehaven the investigator was hosted in large part by a former research director and assistant superintendent of the community. This person retains friendly relations, but no official ties with the present administration. He provided an introduction to the current superintendent of Moosehaven and his staff, and also supplied much information from his own files and important insights based upon his personal knowledge and recollections.

The superintendent of Moosehaven generously cooperated in two long interviews. His assistant and his secretaries were also very helpful. The investigator spoke informally to some 30 to 40 residents during the course of his visit, asking them about themselves and their lives at Moosehaven. A number of visitors, members of the Loyal Order of the Omoose, were also interviewed.

The secondary sources on Moosehaven consisted primarily of the "official" history of the Order as written by Warner Oliver, of back issues of the *Moose Magazine,* and of the annual reports of Moosehaven and the Moosehaven Research Laboratory, to the extent that they are available. All the formal publications of the Moosehaven Research Laboratory and its professional staff, during the years of its operation, were consulted.

One of the major reasons for the choice of Levittown was the availability of an abundance of research data from previous sociological studies of this community. Each of these studies was carefully analyzed for information relevant to our present concerns. Fortunately, these studies collectively spanned nearly the entire history of Levittown over the past 25 years.

These data were supplemented by a variety of materials from the Levittown Public Library, back issues of the *Levittown Tribune,* and the *New York Times.* The investigator spent several days in Levittown interviewing businessmen, a member of the school committee, the chairman of the Chamber of Commerce, two clergymen, members of a youth group, the editor of one of the local newspapers, the community's self-appointed historian, numerous housewives, several realtors, and a representative of the Levitt organization.

The investigator and his wife visited the Bruderhof twice. The first visit provided an opportunity to get generally acquainted and to purchase the several volumes by Emmy and Eberhard Arnold which constitute the primary documentary sources of this case study. The second visit lasted for four days. It provided an opportunity to informally interview dozens of the members, to participate in meals, meetings, and informal gatherings, to work in the communal factory and to peel potatoes in the kitchen. A number of external sources were also drawn upon. Among these are the writings of one of the more youthful ex-members, as well as discussions with several families who had in the past been a part of the Bruderhof.

It is evident that each of these three cases had to be approached differently. The Community Schedule (Appendix B) proved to be a useful unifying link. Despite the considerable differences in approach it was possible, in the judgment of the investigator, to collect data on level II which meaningfully reflect the issues with which we are here concerned. As suggested earlier, data on the family level are, of necessity, less complete and less reliable.

Upon tentative completion of data gathering, an effort was made to respond, in writing, to each of the items contained in the Community Schedule. Where loopholes appeared, additional data were gathered. The descriptive sections of the chapters which follow are based upon the information provided by the Community Schedule. Each of the descriptive reports was mailed to a knowledgeable person within the community under discussion for his comments and corrections, prior to final publication.

CHAPTER 6

# MOOSEHAVEN: AN ADMINISTERED COMMUNITY

Moosehaven is a community of aged and retired members of the Loyal Order of Moose, (L.O.O.M.), a fraternal order. On May 1, 1971, Moosehaven had 465 residents, 62 percent of whom were men. The median age of the residents, though not precisely known, is in the mid-70's. Moosehaven has 168 employees, some of whom reside within the community but are not counted as residents, and most of whom commute from outside. Married couples account for 25 percent of the resident population.

The residents live in houses, each of which accommodates between 25 and 40 persons. Most of them have private rooms, or share with not more than one other person. Each house has its own kitchen and communal dining facilities. Food is reputed to be exceptionally good, by institutional standards. In the newer buildings, toilets and lavatories are semi-private, whereas bathrooms are communal. Each house functions as an operational, administrative and social sub-unit of the community under the direction of a live-in house supervisor.

Residents who become temporarily or chronically ill are housed in the large health center which has a capacity of 150, but is not full. The health center, located within, but at the edge of the nearly 60-acre campus, functions in the manner of a modified nursing home. The discussion of Moosehaven which follows deals only with that part of the population which is not resident in the health center. Moosehaven is a relatively isolated, self-contained community situated some 16 miles from the center of Jacksonville, Florida. Located on an inland waterway, it provides a rural setting in relative isolation. Only in recent years has the surrounding area begun to acquire suburban characteristics. Since residents of Moosehaven do not own private automobiles and since public transportation is sparse, movement in and out of the community is limited.

### History

Moosehaven, established in 1922, is one of the first retiremment communities in America. Its history may best be divided into three periods: the idealistic period (1922-1949), the scientific period (1949-1961), and the institutional period (1961 to the present).

On November 1, 1922, after thirteen years of planning on the part of the Supreme Lodge of the L.O.O.M., the first twenty-two residents came to take possession of an old hotel property and the adjoining estate which has remained to the present day the site of Moosehaven. These twenty-two pioneer residents came as a group—they had previously lived together at Mooseheart, Illinois, the children's village sponsored by the Order. The first residents came with the idea that they would collectively help and sustain each other in their days of retirement. Some would do the cooking, others the cleaning, and others the repairing. The secretary of the Board of Governors of Moosehaven, a person appointed by the Supreme Lodge, resided at Moosehaven during these early days. He appears to have acted somewhat like a general overseer and admissions officer. However the popularly-elected chairman of the Executive Committee, one of the residents, is reported to have been fully in charge of the internal affairs of the community. He was in effect, "a sort of unofficial superintendent."[1]

The historian of the Order reports that the original idea was that "each Moosehaven citizen should contribute all that he could of his material and physical assets to the project and in return he had security for life ... It was almost a Marxian principle."[2] Every resident had a job for which he was paid a nominal sum. Work roles were individually selected and approved by the group. Those who for reasons of health could not work were called "sunshiners." They were paid to smile at passers-by. Originally all wages were equal; however, after the first few months of this experiment, a graduated wage scale based on type of work was introduced.

Information about events at Moosehaven between its first year and the year 1949 is scarce. The community grew slowly. The years of Depres-

[1] Warner Oliver, *Back of the Dream* (New York: Dutton, 1952) p. 181.
[2] *Ibid.,* p. 180.

sion were difficult financially and the community almost did not survive. During this period the position of Superintendent of Moosehaven was formally established. One of the current residents, a woman who has been at Moosehaven since 1939, remembers how, when she was a cook in one of the residences, she and the other cooks used to meet once a week at the home of the Superintendent to plan the meals. It appears that throughout this period, residents continued to play an active role in the management and the maintenance of the community.

At the initiative of the Supreme Lodge, and most especially as a result of the urgings of Martin L. Reymert, Research Director of Mooseheart, a professionally staffed gerontological research laboratory was established at Moosehaven in 1949.

Robert W. Kleemeier, the first Director of the Laboratory, was not only concerned with abstract research, but also sought to directly influence the program of Moosehaven. Kleemeier attempted to strengthen resident participation in the affairs of the community. The individual residences were organized, elections were held, and self-management committees were established within each residence. There were also periodic community-wide meetings. A monthly newsletter was begun, the *Moosehaven Booster,* written by the residents and edited by the research staff. A suggestion box was placed in each residence and subsequently the research staff made careful analyses of the results. A research advisory council was established whose purpose it was to assist the Laboratory in its work.

The Research Laboratory undertook responsibility for the initiation of new residents into the community. It inaugurated a testing program and introduced a formal system of record-keeping. Psychological counseling services were provided. The Laboratory made several detailed studies of the resident population, of death rates, and of turn-over, and introduced standardized measures of physical and mental health.

In the early 1950's, there were 350 residents in Moosehaven and about 60 employees. The mean age of the residents was 76.2 years, and 23.7 percent of the population were women. Of all the male admissions during a 15-month period in 1951-52, 32.8 percent had never been married and 11.2 percent were divorced. This indicates that Moosehaven had an unusual, highly selective retirement population.[3] All residents were white, (as were, and are, all members of the L.O.O.M.), and most were skilled or semi-skilled craftsmen. There were few professional persons. The mean number of years of schooling was 7.7 among the men, and 8.1 among the women. It was found that the average stay at Moosehaven was 7.2 years and that the population had an 11-13 percent turnover annually. In 1951, 60 percent of the population participated in the work program, about half of these being employed in the kitchen and in the dining rooms. The research staff concluded that not only was the work program a significant morale booster, but it obviated the need for some 60 or 70 additional employees.

The Moosehaven Research Laboratory began to fade in the mid-50's, though it did not officially close until the early 1960's. The main causes for this decline appear to have been the death of Dr. Reymert, the gradual loss of interest in research on the part of the Supreme Lodge, and in 1958, the resignation of Dr. Kleemeier. The town meetings and house committees which had been initiated by the Laboratory gradually died out. There was lack of interest in the elections and the last of the committee members had to be appointed by the staff. The suggestion boxes had been discontinued even earlier. The *Moosehaven Booster* continues to this day, under administration auspices. The assistant superintendent, who like the Superintendent is an experienced, dedicated and knowledgeable person, but not a trained psychologist or social worker, now serves as guidance counselor. Difficult cases are referred to outside professionals for help.

In the past twenty years, the resident population of Moosehaven has increased by more than one hundred; the size of the staff has almost trebled. The staff-resident ratio has increased from 1:5 to almost 1:3. Today, 39 percent of the residents are women, an increase of over 15 percentage points since 1950. There are 59 married couples. According to the superintendent, today's population is more dependent than earlier populations: people come to Moosehaven as late as possible in their lives. Today, with greater support from Social Security and with Medicare, and because of the general increase in prosperity, potential residents are able to remain in their home communities longer. They come to Moosehaven only after all other options have been exhausted. For this reason, explains the superintendent, the size of the staff has had to increase and the dependence of the community on the work of its residents has decreased.

---

[3] The population to this day continues to include an extraordinarily large number of isolated men. There are no contemporary statistics which may be compared with these figures from 20 years ago. However indications are that the nature and social class of the resident population has remained fairly constant.

### Moosehaven and the Loyal Order of the Moose

Moosehaven is owned by the L.O.O.M. and largely controlled by its governing body, the Supreme Lodge. The Supreme Lodge appoints a Board of Governors which functions much like the Board of Directors of a traditional voluntary social agency. The membership of the Board of Governors partly overlaps with that of the Supreme Lodge. The Superintendent of Moosehaven is a member of both the Board of Governors and of the Supreme Lodge. He exercises overall operational control of Moosehaven, and appears to serve as the single major influence in the formulation of Board policy.

The Superintendent selects and appoints the staff. The residence supervisors are drawn preferably from among the members of the Order. They ase charged with overseeing the staff within each house (cooks, janitors, dishwashers, etc.) and keeping peace and order among the residents. They are paid employees, not residents. By contrast, in earlier days the cooks within each house served in the capacity of residence supervisors. These cooks were, in most instances, the wives of retired members of the Order. Thus they were more nearly residents than staff.

Staff members below the level of house supervisors are generally not expected to be members of the Order. As often as not they are ineligible for such membership because they are Black.

All residents of Moosehaven (or their husbands) must have been members of the Order for at least fifteen years prior to the date of their admission. Applicants must be at least 65 years old, "be physically and mentally sound, and have a need for Moosehaven service."[4] Chronically ill bed patients are not accepted. However, if a resident becomes chronically ill subsequent to his admission, Moosehaven will care for him. As one resident put it, "This place is paradise for a man who has no place to go." The Superintendent reports that throughout the years, all who are eligible for residence in Moosehaven have been accepted. Moosehaven has, to date, not known overcrowding.

For the residents, Moosehaven is a last home. It provides financial, social, and medical security while aiming to preserve and maximize individual independence and dignity. Residents do not consider themselves the objects of charity, but the just inheritors of entitlements that

---

[4] Carl A. Weis, "Moosehaven Service," *Moosehaven Magazine*, Vol. 57, No. 4, May 1971, p. 19.

have accrued to them by virtue of their membership in the Order. All residents are members of Opportunity Lodge, the local Moosehaven Lodge of the L.O.O.M. Every person retains his own identity. His city of origin and his home Lodge number are inscribed above the door to his room for everyone to see. The rooms are attractive and ample allowance is made for individualization in furnishings and decor.

The L.O.O.M. pays approximately 60 percent of the costs of operation of Moosehaven. It has been the policy of Moosehaven to refuse federal funds for all purposes with the exception of Social Security benefits which are turned over, by prior agreement, by each individual resident. Residents are also individually insured by Medicare; however, the health center is not a federally-approved "extended care facility," and therefore not eligible for Medicare funds.

Moosehaven is a showplace, a mecca of pilgrimage for the members of the Loyal Order of Moose. It and Mooseheart, the children's village in Illinois, constitute the supreme and tangible witness to the charitable and fraternal purposes of the Order. While providing for its dependent members, it serves the entire Order as a symbol, as a rallying point, as a justification for being, and as a source of legitimacy. There are currently some 1,200,000 men and women of the Moose. Few will ever be served directly by Moosehaven, but all contribute toward its support. Especially the Supreme Lodge, the central governing body of the Order, views Moosehaven as a unifying force bringing together the disparate lodges from across the country. The Order held its 1972 national convention at Moosehaven, in honor of its 50th anniversary. The strength and power of the Supreme Lodge is directly dependent upon the size of the membership of the L.O.O.M. Moosehaven has, over the years, become an essential element in the building of this membership.

### Life at Moosehaven

From its earliest days, the governance of the L.O.O.M. has tended to be autocratic. The Directors General throughout the history of the Order have managed its affairs with a firm and controlling hand.[5] This tradition has been carried over to Moosehaven in the role of its Superintendent.

The relationship between the Superintendent and the residents of Moosehaven has its best analogy in the relationship between an employ-

[5] The Director General of the L.O.O.M. functions more or less like the Executive Director of a national agency. The nearly century-old Order has had very few changes of Director General over the years.

er and his employees. The system of relationships appears to have evolved over the past thirty or forty years, and is the consequence of commonsense practicality. The Superintendent or his assistant admits (hires) the residents and he assigns a "home" and a "job" to each. The Superintendent has the authority to fire the workers, but only in extreme cases, such as in the event of severe mental illness, are resident involuntarily removed from the community. Residents receive pay envelopes on payday. They are entitled to one month vacation each year. The vacation may be taken away from Moosehaven with relatives or friends. A person who is dissatisfied with his work, or with his residence, who quarrels with his neighbor, or who has other complaints, turns to the Superintendent for help. "I listen to only one man around here," explained one resident, "Mr. Charles McCall, the Superintendent."[6]

Upon admission to Moosehaven, new residents agree to place into an escrow account with the Supreme Lodge all financial resources which they possess. The administration of Moosehaven may draw upon these funds at the rate of $50 per week. If a resident chooses to leave prior to the time that these resources are depleted, the balance is refunded. However, in practice, most newcomers arrive with little or no savings, and few persons leave the community after they have been residents for a year or more. All current external income of residents is, by agreement, turned over to the community. Monthly income from work assignments at Moosehaven, which ranges from a minimum of $10, to $50, can be spent on beer, tobacco, or postage stamps. "The people all have dollar signs in their eyes," says the Superintendent, by which he means to indiciate that money *is* important to the residents.

There are few other uses for money except for an occasional meal at a restaurant down the road, or for special expenses incurred during vacation. Among residents there are almost no monetary exchanges. The near absence of such horizontal relationships may be symbolic of a larger phenomenon. The visitor senses a lack of depth in the relationships among residents. People are friendly and helpful to each other, but each lives his own life and does not want to get involved with others. This suspicion is reinforced by several anecdotes told by the staff. Some months ago a resident started walking into the river, announcing to two others who were fishing nearby that he was about to commit suicide.

---

[6] The Superintendent of Moosehaven, in reviewing a draft of this text, disagreed with the use of the words, employer, employees, jobs, work, wages, and paid. He would have preferred "duty" for work, "assignments" for jobs, etc.

After having waded into the water up to his knees, the man, calling for help and being ignored, struggled back onto the bank in great disgust and anger. Another man, who frequently came home at night after having too much to drink, was ignored by his neighbors until be began to disturb them with his noise. The problem was brought to the attention of the superintendent who, with the use of a feigned threat, led him to change his ways. In each case it was the staff or the Superintendent, not the individual's peers, who mediated the issue.

"At an earlier time in the history of Moosehaven, relationships among people were warmer and friendlier," remembers an oldtime resident. "The community was smaller and you knew everybody. There weren't any black people here and we were like one big happy family. Things have changed now. Now people only do what they have to do or what they are paid for. In those days people helped each other." This view was corroborated by several residents.

As indicated, there are relatively few families in Moosehaven. There are no children or pets. The role of these two-person families remains an area which is unexplored. At Moosehaven, women have equal rights with men. Whenever husbands and wives are separated due to the transfer of one or the other for health reasons, the marriage partner is given the option of joining his mate in the health center. Frequently the option is not taken and the healthy partner remains alone in the family's room.

Visitors, members of the Loyal Order of the Moose, come daily from throughout the country, proudly admiring the fruits of their charitable endeavors. They appear to be pleased. The residents seem to know intuitively that they are all "shunshiners," smiling so that their fraternal benefactors may be proud and the Order and Moosehaven may be strengthened. Among members of the Order it is known as "The City of Contentment."

### *Analysis*

Moosehaven was selected for study in the expectation that it would serve as an example of an Administered Community, i.e., a community in which the external level and the community level, in partnership, are high goal-oriented and in which the family level is low goal oriented. In this case, the external level, level I, is represented primarily by the L.O.O.M. The community level, level II, is Moosehaven, and the family level, level III, consists of families and individuals who are residents of Moosehaven.

On the basis of the data presented above, it would appear that, especially during the last twenty years, Moosehaven has, as anticipated, approximated an Administered Community. The model leads us to a single out and identify the following social-structural characteristics:

1. High goal-orientation on levels I and II.
2. Contractual partnership between levels I and II.
3. Formal hierarchical structures on levels I and II.
4. Staff oriented towards the solidary goals of the partnership.
5. Staff-resident split.
6. Staff exercising coercive, utilitarian, and normative power.
7. Roles of residents defined by level II.
8. Limited communal organizational subsystems (families).
9. Low goal-orientation on level III.
10. Absence of formal hierarchy on level III.
11. Exercise of normative power only (social control) on level III.

In the case of Moosehaven there is ample evidence both of high goal orientation on the external and the community levels, and of a partnership between these levels. The leadership of the Order, the Supreme Lodge, and the leadership of Moosehaven, the Superintendent and his supervisory staff, are linked in a contractual partnership. In exchange for the financial support provided by the Order, Moosehaven serves as a showplace, a "City of Contentment," enhancing the glory and giving moral purpose to the Order. The specific goal of the Order is the enlargement of its own political and economic power which is most effectively achieved by means of membership growth. The specific goal of Moosehaven is similar, to contribute to the growth of the Order by serving as a happy, successful, attractive witness to the fraternal claims of the Order.

A question may be raised as to whether the specific goal of Moosehaven is, indeed, a specific goal in the Parsonian sense. As such, it should constitute an output into another system and be amenable to contract. It should be measurable; it should be possible to determine when the goal is being approached.

It would appear that the goals of Moosehaven come close to meeting these criteria. If residents were unhappy, the physical plant unattractive, and much dissatisfaction were to emanate from Moosehaven, chances are it would not long continue to receive financial support from the Order. The experience of the short-lived Research Laboratory is a case in point. As originally conceived, the Laboratory was to bring prestige and academic distinction to the Order. With time, it was realized that the kind of prestige that the Laboratory was able to generate was of little significance to the membership. The Laboratory was consequently phased out. Whereas goal orientation is in this case not directly measurable, it might be suggested that the degree of financial support that Moosehaven receives from the Order is a measure of its success.

The position might be taken that Moosehaven's specific goal is to care for the retired members of the Order. This is analogous to saying that the specific goal of General Motors is to produce automobiles, rather than that its specific goal is the production and maximization of profits. This is a defensible position but less convincing than the first. Moosehaven serves only a relatively small proportion of the members of the Order, yet all of them support it. This second goal, service to retired and indigent members, appears, in fact, to be an instrumental goal leading to the accomplishment of the first, the strengthening of the Order. But in either event, considering one or the other to be the specific goal of Moosehaven, the analytical argument does not change. There is a specific goal shared by levels I and II, by the Order and Moosehaven. It is a solidary goal shared in the partnership between the two levels.

Within Moosehaven there is a distinct staff-resident split. The staff is hierarchically organized under the direct control of the superintendent. The superintendent is a part of the leadership subsystem of the most significant element of the external system, the Supreme Lodge. The external system is similarly hierarchical. The primary purpose of the staff is to work toward the attainment of the specific goals of the partnership between the community level and the external level. The residents are only tangentially related to the specific goals of the partnership. Were they more than "sunshiners," were they more highly oriented toward the goal of building the Order, then Moosehaven would more nearly constitute an anti-community: Totalitarianism.

The theoretical model overlooked one distinction which quickly becomes evident at Moosehaven. The staff must be divided into two subcategories: the administrative staff and the maintenance staff. Whereas the most obvious distinction between these categories is that of hierarchical status, there is another, more substantive distinction. The administrative staff must be responsive primarily to the demands of the external system. The maintenance staff, which from the point of view of social status is below the residents, is in a position to be more responsive to the needs and the demands of the residents.[7] The kind of conflict of loyalties that might arise on the part of lower echelon staff when the demands of the administration are in conflict with the preferences of the residents was not observed at Moosehaven. But that does not mean that such stresses do not exist.

The administrative staff (but not the maintenance staff) have the authority to exercise both coercive and utilitarian power. But coercive power is used only rarely in Moosehaven. It is not needed and its exercise might undermine the very goals of the partnership. Thus, few residents are ever involuntarily discharged from Moosehaven. Utilitarian power is exercised with restraint. The wages that are paid to the residents in compensation for their work are of importance to them. The wage differentials between jobs are significant from the point of view of the residents. However, since the basic necessities of life, food, housing, and clothing, are guaranteed, and since everyone is guaranteed a minimum "sunshine" wage, the utilitarian power exercised by the administrative staff has its established limits.

The major controls upon deviant behavior in Moosehaven are normative in nature. But it is important to recall that the residents, in coming to Moosehaven, have agreed not only to surrender their assets, but also to obey a set of norms in exchange for guaranteed security for life. The coercive and utilitarian power of the administrative staff is exercised in the admissions process in a manner which will reduce or minimize the need to exercise anything but normative power at a later stage. The roles of the residents in Moosehaven are largely defined by the administrative staff. In their first weeks at Moosehaven newcomers are expected to learn from their peers the behavior appropriate to a resident. Those who cannot, or choose not to, accept these behavioral norms, leave at their own option. If they are poor, they have no such option and therefore they are forced to conform.

There are families at Moosehaven in the sense that there are married couples. But there are no children. Due to communal eating and housekeeping arrangements, the family unit, while it persists, is of

[7] A similar phenomenon was observed in a state penitentiary by Gresham M. Sykes, *The Society of Captives* (Princeton: Princeton University Press, 1958) p. 40, f. At Moosehaven, residents distinguish between administrative staff and maintenance staff by addressing the first as Mr. or Mrs. and the second by first name.

minimal functional importance. Most of the residents of Moosehaven are not married and there is some evidence that the marital bond, where it exists, is not of the deepest significance. It might be concluded, therefore, that as anticipated, only limited communal organizational subsystems survive within Moosehaven. But this is a moot point. It is possible that these limited family structures result from the extreme age of the population and are not the consequence of the particular social structure which is an Administered Community.

Level III, the residents, are low goal-oriented. However, to the degree that there is a contract between the residents and the staff, it may be said that level III has a specific goal. The contract says, in effect, "we of the administrative staff will guarantee your security for life on the condition that, to the extent of your ability, you will put on a good show for us." That such an implicit contract exists, is beyond doubt. But how pervasive its implications are is open to question. As suggested earlier, were it ever to be adopted as a chief principle of operation, Moosehaven would cease to be an Administered Community and more nearly resemble an anti-community.

In the latter event, if what we have termed decommunization were to take place, we would anticipate the establishment of a more formal hierarchy among residents, together with the exercise of coercive and utilitarian power. Today, one of the striking features at Moosehaven is that among the residents there is, by American standards, an extreme degree of levelling. All residents are considered equal. The norms of the residents require that no person act like a "big shot." If there is a status system among the residents, then it is created primarily by a factor which is beyond their influence and control, the status provided by good health and an alert mind.

The residents have no capacity to exercise either coercive or utilitarian power over each other. This point is dramatized by the fact that, for all practical purposes, there are no monetary exchanges among residents. There is no horizontal economy. There is nothing that individuals can give or take from each other except honor, praise, esteem, ill will, or reprimand—basic elements of normative power.

On the basis of this analysis it may be concluded, therefore, that Moosehaven today largely fits the model of an Administered Community. As indicated, it exhibits a potential leaning in the direction of Totalitarianism. The suspected superficiality in relationships among residents, if it exists, would be supportive of such a trend, were it to take place in the years that lie ahead. It will be recalled that anti-communities are characterized by the absence of horizontal linkage by active generalized cooperation.

Of special interest are the organizational shifts that have taken place over the past fifty years. In its earliest days, Moosehaven began without a staff-resident split, though it must be pointed out that in the person of the Chairman of the Board of Governors, the split existed in vestigial form. The evidence from this early period is sparse, but there are indications that the early Moosehaven of 1922 much resembled an Intentional Community with the external level playing a benign, low-goal-oriented role. In those days a status system among residents had functional significance, self-government was real, not only symbolic or ceremonial as in the Opportunity Lodge of today. It would be tempting to take this analysis further, but available data do not permit it.

Over the years, the community grew in size and gradually a staff-resident split developed. At first, the distinction between staff and residents was somewhat unclear. Individuals moved readily from one status to the other. Perhaps only with the introduction of Black maintenance staff was the full dramatic impact of the staff-resident split felt.[8] Though no historical record has been uncovered of the discussions which took place around this issue, it must have been a very dramatic turning-point for Moosehaven.

The period of the 1950's, the interlude referred to above as the scientific period, is of interest for two reasons: first, because of what the researchers felt needed to be accomplished; and secondly, because of the apparent paucity of their long-range impact. The record shows that the research staff made numerous efforts not only to raise the professional standards of Moosehaven, but also to strengthen the resident participatory structure. Undoubtedly they were aware of the early history of Moosehaven and sought to recapture, by exercising administrative inventiveness, what had been lost. But these efforts quickly fizzled and their long-range effects are almost nil. The net result of the scientific period, in terms of the history of the organizational structure of the community, was, at most, to legitimize the authoritarian rule of the staff. On the other hand, it is likely that a similar degree of legitimacy would

---

[8] It is interesting to note that at Mooseheart, the sister institution of Moosehaven, there is to ths day no Black staff. In that case, whereas all staff are generally members of the Order, the "residents" are children, and as such constitute a distinct social class.

have been achieved by the present administration even without the scientific interlude.

What does this history of organizational changes reveal about long-range shifts in community structure? First, it confirms the fact that changes from one type of community to another do, indeed, take place. But why do they take place? The Moosehaven story suggests that in this case, the reason for organizational changes can in part be related to the expanding demands of the high goal-oriented external level and community level systems. Equally, or perhaps even more important, however, are the changes that have been taking place on the family level. The residents, as a class, have been becoming less healthy, more dependent, and more passive. An Administered Community might never have evolved out of what appears once to have been an Intentional Community had all subsequent residents remained as vigorous as the first pioneers. But in that event it probably would not have remained a retirement community for financially dependent, lower middle class men and women.

There is no evidence that Moosehaven is experiencing democratization, that is evolving in the direction of a Crescive Community. The high goal oriented partnership between the external level and the communityy level exhibits no signs of goal erosion. An anti-community of the type wherein one level is high goal oriented, to the virtual exclusion of the other two levels, is similarly unlikely to evolve because both level I and level II need each other. The basis of their partnership appears to be stable. In short, only an improbable evolution of a Totalitarian anti-community is in sight for Moosehaven. But in all likelihood, Moosehaven will remain the relatively stable Administered Community that it has been over the last twenty or more years for many decades to come.

# CHAPTER 7

# LEVITTOWN, LONG ISLAND: A DESIGNED COMMUNITY

Today there are three Levittowns in America, and of these three the Long Island community is the first. Each of the Levittowns is a suburban community, the product of a single developer-builder who, with aggressive foresight and ingenuity, combined the methods of assembly-line production with vertical integration in business organization to produce bargain houses at bargain prices.

Twenty-five years ago, in May 1947, Levitt and Sons conceived the idea of constructing 2000 rental houses on what was then open land, potato fields in far away Nassau County, 32 miles west of Manhattan. As a result of changes in federal legislation, the plan quickly changed to include an ever-increasing number of houses for sale, and ultimately, the rental houses also were sold. Over a period of four years, even more land was added to the original tract. By February, 1948, there were 6000 houses, and three years later in November, 1951, it was finally announced that the project was complete, with 17,447 houses. "We never had a master plan," admitted Alfred Levitt at the time of completion. "We never knew from one year to the next how much more we could build."[1]

The Levitt houses sold quickly. The original settlers were all World War II veterans. Later, after 1949, non-veterans were also admitted. Three conditions: veteran status, minimum income, and Caucasian racial background, led to the selection of a young, homogenous, first population. The great demand for housing after the war, the trend

[1] Alfred Levitt, "New Towns," *Architectural Forum,* quoted by John T. Liell, *Levittown, A Study in Community Planning and Development,* (Doctoral Dissertation, Yale University, 1952, Ann Arbor: University Microfilms), p. 101.

toward suburbanization, the relative prosperity of the times, favorable federal housing legislation, and the GI Bill were combined by William Levitt to produce the phenomenal financial success which is Levittown.

Levitt built and merchandised houses. Though his advertisements in New York City newspapers invited families to become a part of a planned, garden community, little was planned beyond the layout of the curvilinear streets and houses. A few amenities, such as swimming pools, "village greens," and Levittown Hall, were included in the plan, but there was little concern for community structure or for services. Garbage collection, telephone, electric, and postal services were expected to respond to the fact that people were there. The schools which exploded over a period of 12 years from an enrollment of 40 in 1947, to 18,575 in 1960, were not Levitt's concern.[2] Each of the 17,447 houses had its own septic tank. The fact that the Levittown water system draws upon ground waters which lie directly below these densely clustered sewer systems appears to have been of little concern to Levitt. Levitt was interested in immediate profit, not in long-term investments.

Young families came and settled in Levittown because they were attracted by the house and because they could afford it. For most of them it was a first move out of the city and a first experience at home ownership. John Liell in his 1951 study of 1088 Levittown families found that fifty-eight percent of heads of household commuted to parts of New York City, despite the fact that the travel time was more than an hour in each direction.[3] At first there were no shops or trees, only houses, streets, and mud. To this day , there are no industries of significance within the boundaries of Levittown, a fact that has led to an exceptionally high tax rate.

The high taxes might have been anticipated in 1948, or in 1950, when Levitt bought up most of the land. Levitt not only did not provide space for industrial development, but also his commercial areas were inadequate. Both of these errors must have cost Levitt dearly, and he did not repeat the mistake in his later housing developments. In the early days there was no experienced leadership among the residents with the foresight to anticipate the inevitably escalating tax rate. At a later date,

---

[2] Levitt sold plots of land at cost to the local school authorities and left it to them to construct the schools. The history of Union Free School District #5 is reviewed in *Levittown, New York, A Study of Leadership Problems in a Rapidly Developed Community*, (Washington, National Education Association, January 1962), pp. 11-13.

[3] *Ibid.*, p. 233.

large corporate entrepreneurs from outside the community discovered the missed opportunities and established shopping centers and commercial enterprises. Major industrial development, which spread rapidly in other parts of Nassau County, has never taken root in Levittown.

In the early days William Levitt attempted to exercise personal control over seemingly unimportant areas of life among the residents. Levittowners were not permitted to put up fences. Laundry could not be hung up outdoors on weekends, and never on long lines, only on circular racks. When the Committee to End Discrimination sought to use the South Village Green for an inter-racial square dance, Levitt refused to grant permission. When the local newspaper, *The Levittown Tribune,* criticized some of his actions, he bought it and began publishing his own views.

Mr. Levitt is aledged to have said at one time, "the masses are asses."[4] There is no evidence that the residents of Levittown reciprocated with their own expressions of contempt.

It appears, however, that after all the houses were sold, Levitt quickly relinquished all control and personal interest in Levittown. The village greens, the swimming pools, and Levittown Hall were all transferred to the town (Hempstead) Park Department. *The Levittown Tribune* was sold. The 25th anniversary celebration of Levittown, which took place in the early part of 1971, was sponsored not by Levitt and Sons, as might have been expected, but by a handful of old time residents. The turn-out at the celebration is reported to have been poor.

There are many ways in which the boundaries of Levittown may be defined. The U.S. Census Bureau uses Union Free School District No. 5, and calls it Levittown. But this excludes many Levitt houses and includes some others which were not built by Levitt. The area in which Levitt built his houses encompasses part of four school districts, four fire districts, four water districts, 2 congressional districts, 2 townships, 5 postal districts and one county. Dobriner counted a total of 21 separate political jurisdictions.[5] One of the residents commented, "I have a Westbury postal address, I'm in the East Meadow School District, the Hicksville Water District, the town of Hempstead, and I have a Levitt House."[6] Periodically the question of independent incorporation for

[4] Quoted by John Liell, *op. cit.,* p. 135, from *Time,* July 30, 1950.
[5] William M. Dobriner, *Class in Suburbia* (Englewood Cliffs, N.J.: Prentice-Hall, 1963) p. 89.
[6] "Levittown Confounds Forecasters of a Slum Future," *The New York Times,* May 11, 1965, p. 41:2.

Levittown has been raised, but never with much persuasiveness. The President of the Chamber of Commerce explained that it would be too expensive, and only increase the excessively high tax rate.

In the early 1960's, when the community suddenly discovered that a tidal wave of children was overwhelming the school system, the financial strains placed upon homeowners led them to engage in a conflict characterized by unprecedented bitterness and strife. Levittown was split in half between those who sought to "improve" the schools and those who wanted to economize. Studies of this period show that the conflict was largely, but not entirely, along social class and religious lines.[7] For a number of years elections for the District #5 School Board were heatedly contested. The pendulum of public opinion swung rapidly from one side to the other. School superintendents were hired, and fired, and hired again. School board meetings often attracted hundreds of excited onlookers and sometimes lasted throughout the night, until 6 A.M.[8]

In an attempt to help resolve these conflicts, both the National Education Association and the State Education Department entered the picture. Orzack and Sanders, two sociologists hired by the NEA, conducted a leadership survey in 1961, and, among other things, concluded that the crisis had been precipitated by, "the total lack, except for the school district, of major centers of political identification, (in which) . . . citizens could exercise some control and to which they could be loyal."[9] They criticized the lack of voluntary organizations except those "which drew upon limited, highly personal and expressive desires" and the lack of mechanisms for sustained coordination. Finally, they criticized the "value system of the community's residents with its stress on home, family, and children" to the neglect of, "education in any broad sense." "Put simply, the fabric of Levittown has little texture. Person-centered activities predominate. Family and home attract much attention. There is a traditional sense of privacy, concealment, and some withdrawal about such personal matters which helps justify separate interests."[10] In their analysis of leadership patterns, Orzack and Sanders found that, "community leadership roles that are stable in duration and variable in sphere of influence have not developed."[11] "Apparently no

---

[7] Louis H. Orzack and Irwin T. Sanders, *A Social Profile of Levittown*, (Boston University, Dept. of Sociology and Anthropology, May 1961, Reprinted, Ann Arbor, University Microfilms; and Dobriner, *op. cit.*).
[8] See, *Levittown, New York*, National Education Association, *op. cit.*
[9] Orzack and Sanders, *op. cit.*, pp. 40-41.
[10] *Ibid.*, p. 38.
[11] *loc. cit.*

organization has developed the exclusiveness which makes membership in it a badge of high social status."[12]

The crisis lasted several years. While the NEA conducted its study, the New York State Commission of Education ordered his own study through the Office of Field Services of Cornell University.[13] When in 1962 the results of this second study were made public it was critically received by Levittowners. The study concluded, like the previous one, that Levittown is "under-organized" and therefore school matters have become the major area for residents to "let off steam." Similarly, the study noted that Levittown had, "a disproportionate number of people who lack political experience and many of the political skills necessary for moderated political action at the community level." A final "philosophical suggestion" proposed by the study was that, "there might be a need to provide stability and control through the state government during a period of transition, and return control to the community on a gradual basis, as they mature and are able to assume the responsibilities which are theirs."

When today's residents and former activists in the school struggle are asked about the crisis of ten years ago, they report that the "liberals" finally won, but without declaring a victory. In recent years, they say, there have been no issues of similar importance and usually school board elections have gone uncontested. "We were young and inexperienced then. Today we don't get excited like that anymore," commented one former school board member.

There are still issues in the community. The 1971 school year began with a three-day teachers' strike. In the Island Trees School District (which covers a part of Levittown) the question of sex education in the high school recently became a matter of major public debate. Several years ago much excitement was aroused by a school teacher who insisted upon wearing a black armband, in protest of U.S. military action in Vietnam. But none of these issues has been as convulsive as the conflicts which occurred in the early 1960's.

### The People of Levittown

The original population of Levittown was uniformly young, married, and with young children. Liell's 1951 representative sample of 1088 families consisted of 62 percent white collar heads of household. The

[12] *Ibid.,* p. 33.
[13] The study was conducted by Dr. Claude L. Kulp. The discussion which follows is based upon the summary of the 437 page report in the *New York Times,* October 19, 1962, p. 33:1.

United States Census for 1960 reported that 51.7% of employed males were white collar workers. Dobriner predicted in 1963 that during the 1960's Levittown would become an increasingly homogenous, lower middle class community. He was only partially correct. The most recent census (1970) shows only a slight decline in male white collar occupations to 48.7%. There is a large increase in service workers, including many policemen and firemen. The number of employed women has nearly doubled over the ten year period.

After all these years, Levittown retains a stigma in the eyes of many upwardly mobile individuals. Some people still prefer to conceal the fact that they live in Levittown by defining their residence as being located in Island Trees, Wantaugh, etc. It is predominantly Catholic, and conservative Republican. In the 1970 United States senatorial race,

**TABLE 1**

*AGE     DISTRIBUTION     OF     POPULATION     IN LEVITTOWN, N.Y., 1950, 1960, 1970*

|  | Year | | |
|---|---|---|---|
| *Age* | *1950\** | *1960* | *1970* |
| Under 5 | 29.6% | 15.5% | 8.7% |
| 5-9 | 11.4 | 16.1 | 11.8 |
| 10-14 | 2.3 | 12.2 | 13.0 |
| 15-19 | 0.8 | 5.3 | 11.4 |
| 20-24 | 4.1 | 2.6 | 6.0 |
| 25-29 | 19.6 | 5.9 | |
| 30-34 | 18.9 | 9.9 | 11.0 |
| 35-39 | 7.9 | 10.9 | |
| 40-44 | 2.7 | 8.4 | 13.9 |
| 45-54 | 1.3 | 7.3 | 14.4 |
| 55-64 | 0.7 | 2.7 | 5.7 |
| 65 and over | 0.8 | 2.8 | 3.6 |
| TOTAL (N) | 1088\* | 65,276 | 65,440 |

\*Based on a sample of 1088 families studied by Liell

*Sources:*

John T. Liell, *Levittown, A Study of Community Plans and Development,* Yale University, Dept. of Sociology, Doctoral Dissertation, 1951, p.212, U. S. Census of Population, General Population Characteristics 1970, Advance Report, PC (v2)-34.

Levittown voted more than 60 percent for the Conservative candidate, in a field of three.

Over the years the population has aged and the children have grown up (see Table 1). Perhaps the unusual age distribution among Levittown residents, more than any other single factor, helps to explain the school crisis of the early 60's and its subsequent abatement. Now maximum enrollment in the schools seems to have been reached.

Most of the heads of household continue to commute in the direction of New York City, as in earlier days. Whereas in the past few women were employed, today, since most of the children are older, a larger number of mothers have joined the labor force and therefore the total income of families tends to be higher. Levittown has few families receiving public assistance, and few persons above the age of 65. To this day, the population is almost exclusively white, despite the fact that all legal barriers to racial integration have been eliminated.[14]

Since many of Levittown's workers are directly dependent upon defense related employment, the recent cut-back in this kind of work has led to a relatively high rate of unemployment. There are small signs of economic distress. In 1971, for the first time in ten years, according to the *Levittown Tribune,* a school budget was voted down by the local electorate.[15]

"This is not a town; it has no heart; it's just a bunch of people who live in houses," explained one of the old timers who has lived in Levittown for over 20 years. "Levittown represents a horizontal version of the vertical apartment living patterns of our urban centers," commented a sociologist in the mid-fifties.[16] Whereas both of these comments are likely to be exaggerations, it is true that Levittown, to the extent that it is identifiable as a separate entity, lacks a political, economic, or physical center. It is highly dependent upon external systems for its survival. Levittown is a bedroom community. Its stores and banks are branches of large chains and corporations whose executive offices are located elsewhere. Issues of local government are decided in the Hempstead

---

[14] In 1970 there were 44 blacks in Levittown and 268 other non-whites out of a total population of 65,440, according to the U.S. Census of Population.

[15] *Levittown Tribune,* August 19, 1971.

[16] Harold L. Wattel, "Levittown: A Suburban Community," in William M. Dobriner, *The Suburban Community* (New York: Putnam & Sons, 1958) p. 299.

[17] The population of Levittown constitutes 8.1 percent of the population of Hempstead. Levittowners elect only one representative of the Hempstead Town Council.

Town Council, a political body in which the residents of Levittown have only a slight, minority interest.[17] There are the Rotary, the Lions, the Kiwanis, and the churches, but as is common in America, all of these associations are local branches of large, national organizations. There is no United Fund and no Community Council. The Chamber of Commerce, which was organized about 12 years ago, is presided over by a businessman who is a non-resident.

There are two Roman Catholic Churches within Levittown, i.e., among the Levitt houses. Parish boundaries not only divide the community, but also include others who do not live in what is generally considered to be Levittown.

Levittown has sixty Boy Scout Troops, twenty-five Girl Scout Troops, and claims to have the largest Little Leagues in the United States. Teenaged youth are scattered among several high schools. High school sports for boys are important. But many other extra-curricular activities of District #5 high schools were eliminated in 1971 as a consequence of the new austerity budget.

To the extent that there is a "sense of community," it appears to exist primarily among women who meet at the grocery store or in front of the elementary school, and somewhat differently among businessmen who are associated in the Chamber of Commerce. Veterans' groups stage periodic celebrations and PTA's hold meetings. There are uncounted clubs and associations, but most of these are small, involving only a few active participants and having no goals that could arouse larger groups to united action around community issues.

There is one notable and important exception to this near absence of community concern. The Levittown library is a comprehensive, well-stocked and staffed community facility located in an attractive modern building among beautiful grounds, just off the main throughfare, Hempstead Blvd. The library is supported by public funds voted annually by the residents of School District #5. It has been generously supported by the people and stands out in stark contrast to the many other community-supported services and facilities which might be, but are not.

Whereas there is little concern for collective community action, Levittowners are extremely proud of their homes which, like automobiles, come in model years. Their homes appear to serve as important extensions and expressions of their personalities. There seems to be no end to the amount of tinkering, repairing, remodeling, and expanding that Levittowners will superimpose upon the original Levitt frame. Today, every house looks different and the mass-produced monotomy is a matter of the distant past. Neighbors appear to demand not conformity, but standards of cleanliness and propriety in the outward appearance of the houses on their street. Contrary to many forecasts of the early 50's, Levittown after 25 years has not turned into a suburban slum.

Now the children of the original settlers are young adults. Many have joined the military and the local paper is full of proud stories of their exploits. Many others are enrolled in Nassau Community College and a small, but outspoken contingent shows signs of identifying with the counter-culture. Today, Levittown has its own "headshop." Young people gather in an informal, store-front youth center where the dress and the conversation, from the point of view of the parents, is far from proper. They are called radicals, revolutionaries, and Communists, but they seem hardly to shrink under these curses which, to them, are empty. One youth proudly announced that he had written his senior high school essay on the theme of racial minorities in America. It appears that at least one segment of Levittown's younger generation is less than satisfied with the *status quo* which represents the pride of their parents' achievement.

### Analysis

In selecting Levittown for study, our purpose was to examine the organizational structure of a community which was expected to resemble the model of the Designed Community. Theoretically, the Designed Community is high goal oriented on the external and the family level and low goal oriented on the community level. Specifically, we expect to discover the following characteristics:

1. Contractual partnership between levels I and III.

2. Reciprocal goals on levels I and III.

3. Formal leadership hierarchy on level I.

4. Staff-resident split.

5. Staff oriented toward level I goals.

6. Exercise of coercive, utilitarian, and normative power by staff.

7. Level II defined by levels I and III.

8. Low goal orientation on level II.

9. Absence of hierarchies on level II.

10. Exercise of normative power only on level II.

11. Vertical linkages of level II by passive generalized cooperation; no vertical contracts.

In applying the above paradigm to Levittown, we must, throughout our discussion, be clear to which period in the history of Levittown reference is being made. In order to avoid misunderstanding, we shall refer to the first five years of Levittown (1947-1952) as the Levitt period. During these years the houses were built and sold, and Levitt and Sons exercised a relatively high degree of control over the life of the community. The subsequent period, after Levitt and Sons had relinquished control, shall be called the post-Levitt period.

The specific goal of Levitt and Sons was to build houses and thereby to garner a profit. Levitt's primary efforts were directed toward the maximization of his profits. This goal was achieved by means of satisfying the goals of the young veterans and their families who were in search of housing and of a new life-style at a distance from the crowded city. A contract, a sales agreement between Levitt and the residents, formalized the link between levels I and III. The goals of the Levitt organization and the residents were reciprocal and, as such, constituted the basis of the high goal oriented partnership suggested by the model.

As indicated in Chapter 5, it is always difficult to assess the nature of level III. In this case, level III *appears to be* high goal oriented. Especially during the Levitt period, families moved to Levittown primarily because of the attractiveness of the contract, i.e., the house was a good buy, a valued package offering status, utility and pride. Many comforts, perhaps even necessities, were sacrificed to achieve this goal. Family, friends, the many informal ties with the past, were relinquished by the newcomers in the pursuit of this goal. Husbands passively agreed to travel many hours to their place of employment in order that their families might benefit from the new home in the suburbs. The singlemindedness with which they pursued this end led them to overlook the fact that the new community was largely unprepared to provide essential services and amenities for the new residents.

During the Levitt period, the Levitt corporation functioned efficiently to achieve its goals. The staff of the corporation, representatives of the external level, controlled not simply the physical development of the community, but to a degree also sought to influence the behavior of the residents. The purpose of all these efforts was to enhance the desirability and the salability of the new houses which were continually being added.

Almost as soon as Levitt had accomplished his purpose, he withdrew from the scene.[18] In the early days, the Levitt organization had minimally

---

[18] Levittowners speak of Levitt as a person, rather than of Levitt and Sons, a corporation, which is today a subsidiary of ITT. Even 25 years later the contract between builder and buyer retains some expressive overtones.

responded to the need of the residents for services by negotiating in their behalf with public and semi-public authorities. But when it departed, many unresolved problems remained. Utilities, educational facilities, problems of governmental organization were dumped into the laps of the unprepared residents or upon other, external organizations and institutions.

There is little evidence that during the Levitt period the Levitt organization, representing level I, used coercion. Occasionally, families must have suffered eviction due to their failure to keep up with their mortgages. The utilization of the power of the state to exclude racial minorities - until it was later declared illegal - serves as an example of the utilization of coercive power. Most commonly, however, the Levitt organization exercised utilitarian power. Families got what they paid for, no more and no less. Projects that Levitt supported benefitted from his "largesse," as for example, Levittown Hall. When Levitt disagreed, as with the *Levittown Tribune,* then he exercised negative sanctions. Most importantly, Levitt was in a position to control who would more into Levittown. Whereas he had only a minor interest in directly exercising normative power (influence) within the community, the net effect of his sales policy was to select a normatively highly homogeneous population.

As indicated, Levitt was not primarily interested in building a community, only in building houses. The absence of organizational development on the community level which is characteristic of Levittown to the present day, is partially attributable of Mr. Levitt's non-plan from the beginning. At an early stage he decided that it was not necessary for the accomplishment of his goal. Subsequently, during the post-Levitt period, for different reasons, the residents chose not to alter this early, weak community level design. In part, this appears to have been what they wanted, and in part, they had no capacity to make it otherwise. The community level, level II, is and always has been low goal oriented. It contains few formal hierarchies and, with the exception of the school system, few formal leaders. There is no capacity to exercise more than normative sanctions over the behavior of individuals and families, and little power to influence external systems. With the exception of the school district and the library, the community level exists only as a generalized idea in the minds of residents and outsiders, with indefinite physical and conceptual boundaries. Only the school system and the library have the capacity to enter into contracts which bind the entire community, with families on the one hand, and external systems on the other.

New families moving into Levittown tend to relinquish, or significantly reduce in importance, most of their former links with systems external to the family, with one notable exception: the link to their employers. This link constitutes the second major dimension of level I during the Levitt period. With the shift into the post-Levitt period it did not change. It is almost banal to say that the source of economic livelihood of the residents still has a profound effect upon the lives of residents of Levittown and that this source remains almost exclusively external to the community.

The interesting questions in the case of Levittown arise in the analysis of its shift from the Levitt to the post-Levitt period. Did the nature of the external level change? Did the external level gradually become low goal oriented? What happened to the contract between level I and III? Did level III, the family level, become low goal oriented?

These are difficult questions to answer because they are largely related to one's interpretation of modern American society. Levittown is almost a caricature of the American community subsequent to what Warren has called "the Great Change."[19] A multiplicity of external systems exercise control over the affairs of the community, of Levittown even more than of most communities due to the near absence of its level II. Had the State Education Department acted upon its threat to take over the school system in 1962, the emasculation of Levittown would have been nearly complete.

Subsequent to the departure of Levitt, the external systems which had previously been less prominent, or which had previously related to the community via the Levitt organization, moved into the foreground. Levitt's own formal contracts had been transferred to the banks. The town of Hempstead assumed partial, though not complete, authority over governmental services. (Only education and fire protection were left to local control.) A number of national organizations, especially chain stores, assumed control of the local commercial outlets. The growth and the influence of the sum of these external systems has been only very partially examined in this analysis. But at this point in history there can be little question that they constitute a very major input into the lives of families and individuals in Levittown. In short, the partnership, if indeed there is such, now exists between level III and a new level I, represented by a multiplicity of national and regional seemingly high goal oriented systems.

---

[19] Warren, *The Community in America, op. cit.*

The Levitt staff departed with the Levitt organization, but it has been replaced by a multiplicity of individuals who are oriented toward the goals of external systems. Thus, the original staff-resident split has been replaced by a new staff-resident split.[20] In short, there is little evidence of what we have defined as democratization. Levittown has not become a Crescive Community.

It is possible, therefore, to take the position that Levittown has remained a Designed Community, the only significant change having taken place in the personnel and the nature of the integration of the external level. In the past, the external level consisted primarily of a single, high goal-oriented system, whereas now it has become fragmented.

But the alternative position may also be taken, that Levittown has experienced a degree of decommunization, that it has moved in the direction of the Solipsistic anti-community. There is some evidence that the external level, because it has become increasingly distant, fragmented, ill defined and arbitrary from the point of view of the family level, has forced individuals and families to become more isolated. Families may be viewed as high goal oriented, struggling to survive on their own, without the support of either community level or external level systems. It is the condition of modern American life which Philip Slater has called "the pursuit of loneliness."[21] Whether, and to what extent it applies to Levittown cannot be answered with any degree of certainty.

---

[20] More accurately, it might be said that this is a new and complex multiplicity of fragmented staff-resident splits.

[21] Philip E. Slater, *The Pursuit of Loneliness,* (Boston: Beacon Press, 1970.).

CHAPTER 8


# THE BRUDERHOF: AN INTENTIONAL COMMUNITY


The Bruderhof, or the Society of Brothers as it has become known in America, is a religious community of some 800 members living in families at three separate rural locations: in Pennsylvania, in New York State, and in Connecticut. Approximately one-third of the members reside at each of the three locations; however, the Brothers consider themselves all one single, spiritually united community.

The central, most important fact about the Bruderhof, from the point of view of the Brothers, is that it is a community unified by love and Christian faith. The church-community (*Gemeinde*) as it is most often called by Eberhard Arnold, its principal founder, is a brotherhood in radical discipleship of Christ in emulation of the primitive Christian communities of the first and second centuries. It seeks to embody a total commitment to a way of life that gives the fullest possible expression to the unity and love ordained by the Sermon on the Mount. Jesus is experienced as love, "as love without violence, as love without rights and without the will to possess."[1] It is based not only (or merely) upon the emotional and the intellectual bonds which exist among men, but what is more important, upon the unity which is made possible by God's Holy Spirit. "The collective soul of the community is the Holy Spirit. In this Spirit the church-community is united."[2] The church-community is brought together and held together by a unity which is greater than, and which transcends itself.

---

[1] Eberhard Arnold, *Why we live in Community* (Rifton, N.Y.: Plough Publishing, 1967) p. 9. Originally published in German in 1927.
[2] Eberhard Arnold, *A Testimony of Church-Community from his Life and Writings* (Rifton: Plough Publishing, 1964) p. 21.

"All that we do could be the expression of unity; but that is only the expression ... The expression perishes, but the unity endures."[3] The Bruderhof is, "a monument in real life by which men could recognize the cause for which Jesus died."[4] Its goal is not limited to the salvation of individuals but to serve as a witness, to prepare the way for God's kingdom on earth (and not in heaven). "Through Christ the kingdom of the future takes place now in the church-community."[5] The fact that for the past 1800 years Christianity has emphasized the redemption of the individual, rather than the kingdom of God on earth, is a "grandiose misunderstanding."[6]

It is only after this "radical anarchism of faith responsible to God alone" is understood as the foundation of all that is Bruderhof, that it becomes meaningful to delineate the form, the lifestyle, and the history of the Society of the Brothers.[7] For all things are logically predicated upon this faith, with almost Euclidean rationality. Without an understanding of it, it may be possible to anatomically dissect the body of the Bruderhof, but it would only be a corpse, a body without life or soul.

### History

The Bruderhof traces its origin to Germany, to the period during, and immediately following the first World War, a period when a multiplicity of religious, socialist and pacifist ideas and ideals were being given excited and practical expression, especially among youth and among the returning veterans. In its search for a more simple life, for new meanings and goals, the period bears striking similarity to the spirit among university youth that prevails in our own times, in post-Vietnam America.

In 1920, Eberhard Arnold, at that time a young theologian, teacher, and national chairman of the Student Christian Movement, together with his family and close friends, established the first Bruderhof in Sannerz in central Germany. Without practical plans, lacking funds or pledges of support, a large house was rented in the countryside to serve as a home for those who would join this young, idealistic, fledgling congregation. They promised complete loyalty and love to God and to

---

[3] Emmy Arnold, *Torches Together* (Rifton: Plough Publishing, 1964) p. 158.
[4] Eberhard Arnold, *Salt and Light* (Rifton: Plough Publishing, 1967) p. XII.
[5] Eberhard Arnold, *A Testimony, op. cit.*, p. 24.
[6] *Ibid.*, p. 17.
[7] Emmy Arnold, *op. cit.*, p. 27.

each other, sharing all their meager material possessions in common. "We were determined to burn all of our bridges behind us and put our trust entirely in God," reminisced Emmy Arnold, Eberhard's wife, some forty years later.[8]

During the first summer, the little community had over 2000 guests, some staying only a few hours, others remaining several weeks or even months. Despite severe financial struggles, by the winter of 1921, some 60 people had been gathered into the household at Sannerz. Some of these were children who had simply been abandoned at the door. Others were young single and married people coming from a large variety of backgrounds. The community was wide open, almost all who came were welcomed into the fellowship.

In 1922, the first great crisis arose. In that year the vast majority of those who had been attracted to the idealistic little community departed. Their argument was that, "people with the new vision should now turn back to the old conditions of life and be a small light there."[9] But the handful that remained discerned within this argument a lack of faith and a selfish desire for return to the material comforts of middle-class life. Those who departed contended that faith and economic matters do not belong together. Eberhard Arnold and those who chose to remain with him strongly disagreed. They insisted that this agrument was the same as that which had led to the downfall of the world-church. They demanded that faith should penetrate and master everything, including financial matters.[10]

Thus in 1922, a new beginning was made with a more highly crystallized philosophical stance. By 1926 the community had again grown to forty or fifty members. In addition to the publishing of religious books, the community had begun to serve as a school for its own children. The meager income of the community was derived from farming, handicrafts, the sale of publications and occasional gifts from friends and supporters.

In 1926 the Bruderhof moved to a new location which became known as the Rhon Bruderhof. This second home served the community for over ten years, until it was finally closed by the Nazis in 1937. The

---

[8] Emmy Arnold, *op. cit.*, pp. 41-42.
[9] *Ibid.*, p. 70.
[10] *Ibid.*, p. 75. By the world-church the Brothers mean the institutionalized, historial Christian churches. By contrast, the Church which becomes immanent within the church-community is the embodiment of the Holy Spirit.

intervening years were times of rapid growth. By 1933, the group had grown to 180. The following year a branch settlement was established in Lichtenstein and a little later, a beginning was made in England. During the years 1930-1931, Eberhard Arnold made a trip to America, visiting with the Hutterites in the United States and Canada. For him it was, in part, a journey in search if identity—a desire to find an anchor among the historical heirs of the sixteenth century Anabaptist movement. As a second purpose, Eberhard Arnold sought financial support. An informal, cooperative relationship between the Hutterites and the Bruderhof was established as a result of this trip, a loose relationship which has continued to the present day.

In 1935 Eberhard Arnold died, leaving behind him a message and an approach to life which has served the Bruderhof as a source of inspiration and guidance throughout the years of its existence. Upon leaving Germany, the community moved first to Lichtenstein, and then to England. In England the numbers swelled to some three hundred. When England entered the Second World War the community, since it contained a large number of "enemy aliens," was again forced to wander, this time to Paraguay. For nearly twenty years the Bruderhof survived in the rural, semi-tropical isolation of this Latin Amiercan wilderness. Subsisting primarily on agriculture, being both physically and culturally cut off from familiar environments, they found life harsh, and thus when the opportunity arose in the mid 1950's the community migrated to the United States. In this country, especially since the establishment of Community Playthings, the toy industry which economically sustains the Bruderhof in the present day, the days of extreme privation have passed. In 1971, after much thought and discussion, the Bruderhof, while retaining its base in the United States, began to re-establish a community in England. Today, after fifty years of wandering and struggle, the Bruderhof is economically and numerically strong, and expanding.

### Life at the Bruderhof

Holding "all things in common," like the Apostles in the New Testament, the Bruderhof organization bears a striking resemblance with the modern Israeli Kibbutz. It is a comparison of which the Brothers are well aware, but it holds little deeper significance. Most meals are eaten in the communal dining hall. In its private residence, each family retains only the small necessities of daily existence. There are no formal exchanges of money or property among the members since all material wealth belongs to and is shared by all. One of the most important organizational differences between Kibbutz and Bruderhof lies in the fact that children, though they spend most of their waking hours in the children's houses or in school among their peers, sleep at home with their families.[11]

In addition to the toy factory which is located on the premises, the community maintains a large garden producing fruit and vegetables primarily for home consumption. A few members are employed by the publishing enterprise with is primarily an educational, rather than a business venture. Many of the Brothers and Sisters are occupied with the children, in preschool and school activities which extend through the eighth grade. Thereafter children proceed to the regional public high school and then to college or professional school. It is thought important that young people, as they grow up, have an opportunity to examine the outside world before they return, if they do, to become Brothers in their own right, and by their own decision.

The Norfolk, Connecticut Bruderhof, Evergreen, as it is known among the Brothers, is situated on a 40-acre rural estate. Some ten years ago the community purchased an abandoned millionaire's "castle" and refurnished it to serve it own needs. The outbuildings have been converted to serve as school buildings and residences. A new factory building, a kitchen-dining room and several multi-family houses have been added in recent years. At present, there are some 280 persons living in this "household," about half of them children. Cleanliness, orderliness, and beauty prevade the surroundings. In the summer flowers and greenery are everywhere. Thoughtful design and artistic creativity have been given full rein in the planning of this environment.

It is toward the end of uniting all things that the community and all its members strive. Thus there is no dividing line between the secular and the holy, the ideals and aspirations of the individual and of the group, work and worship, one person's joy and another's pleasure. Only inevitable human frailties, selfishness and lack of openness and honesty, are seen to set limits upon the capacity of the community to achieve its aim.

---

[11] In recent years, a number of Israeli Kibbutzim have changed with respect to children's houses. These few Kibbutzim are similar to the Bruderhof in their organization in that children sleep in the home of the parents, rather than in the children's houses.

The Sermon on the Mount, the historical example of the early Christian communities and of the Anabaptists, and the writings of Eberhard Arnold serve as major guideposts in the life of the community. There is no theological hair-splitting caused by doctrinal differences. The basic principles are few and simple. Children are taught not by the verbal inculcation of doctrine, but primarily by the living example of of their elders.

The Brothers meet as often as possible, four, five, or even six times a week, in order to re-establish and reconfirm their unity withiin the Holy Spirit. It is their joy to be together as often as possible. "The Bruderhof is not Church. . . . The Bruderhof *becomes* Church whenever the Holy Spirit, the Jerusalem above, comes down upon us; whenever the Church unites us in the Holy Spirit."[12]

The Brotherhood meetings are restricted to the Brothers and Sisters who have fully committed themselves to Christ and to life in the church-community, plus the novices. The meetings are conducted somewhat in the style of Quaker meetings, each person, man or woman, expressing himself as the spirit moves him. The most important decisions affecting the community are made in this setting. Discussions may range from the mundane and practical, to the interpersonal, to the spiritual. As more than one Brother explained, when people really love each other, when they live together as in one big family, then they also have great capacity to hurt each other. That is why it is important to meet together constantly and to re-establish unity.

In principle, decision-making is by consensus. Only routine, practical, or purely personal decisions are made by individuals. All other matters are brought to the group. "When a community of moved people believes in the Spirit, the freedom of the individual lives in the free decision of the common will brought about by the Spirit. Freedom, as the will of the Good, brings about unity and the unanimity from within."[13]

In addition to the Brotherhood meetings, there are Household meetings which are held less frequently and which include the younger members of the household, not only the Brothers. The *"Gemeindestunde"* which is usually, though not necessarily, held on Sunday, is the only meeting which has primarily worship as its purpose. But it is not to be

---

[12] Eberhard Arnold, *Love and Marriage in the Spirit* (Rifton, N.Y.: Plough Publishing Co., 1965) p. 38.
[13] Eberhard Arnold, *Why We Live in Community, op. cit.,* p. 15.

compared with a church service, for it is not in any way special or more important than any of the other meetings.

Individual Brothers are selected by the community to fulfill leadership functions such as Men's Work Organizer, Women's Work Organizer, Shop Superintendent, etc., but "there is no difference at all in worth; only a difference in calling."[14] The emphasis upon equality informs each person to exercise his responsibility with quiet humility. Again, when human failure leads to ambition and arrogance, then the Brotherhood must support the individual to discover his error. Usually, persons in leadership positions retain them indefinitely, until they ask to be relieved of the responsibility.

The Servant of the Word is an individual—usually at any given time several individuals—who is responsible for the "inner life" of the community. His role is one of arbitrator, consoler, helper, coordinator and inspirer. "He who has this task must first grasp the inward and outward situation of the whole group and bring it to clear expression in word and action. He must bring to expression that which is holy, which moves and fills the hearts of all, even if it is unspoken and undone."[15] The Servant of the Word is not viewed as one who acts as an individual, but rather one who gives expression to the immanent consensus which resides in the total community.

The enhancement of individual talent, when not committed to the service of others or of the community as a whole, is discouraged. Self-centeredness, whether it is given expression by means of the accumulation of excessive personal possessions, or talent, or knowledge is viewed as a source of disunity. This does not, however, prohibit individual differences of taste and preference. "Community lives only in living reciprocity. Therefore, rejoice in your diversity and never be offended by it!", taught Eberhard Arnold.[16]

The Bruderhof gives the impression of a well-coordinated, efficient, sensibly organized society in microcosm. The communal meals begin punctually. First there is an extended silence and then a song; another silence and then eating and soft conversation begins. Often the time of eating is accompanied by reading from a novel or a short story, or by a concert. The visitor is surprised that meals neither begin nor end with a

---

[14] Eberhard Arnold, *Love and Marriage, op. cit.,* p. 31.
[15] *Ibid.,* p. 237.
[16] Emmy Arnold, *op. cit.,* p. 142.

prayer. It is the meal itself that is a love-feast, a religious act, a reenactment of the Last Supper. There are few, if any, explicitly religious rituals in the life of the community. With each special occasion, be it a birthday, a wedding, or the celebration of a holiday, the community considers anew how it might observe this event. Singing, dancing, play-acting, laughing, joking, eating sweets and smoking, playing games, these people are no ascetics.

Generally, people are at work six days a week, eight hours a day. Women work slightly shorter hours than men. Work and rest periods are punctuated by the sounding of a bell. Schedules vary, of course, for those who are occupied with the children or in the kitchen. Each person is fully trusted to do that of which he is capable, and to do it well. While competition is discouraged, it is most likely to occur among those who want to do more, rather than less: "Justice does not live in harsh demands upon others, but in the joyful sacrifice of what is one's own."[17]

In a purely technical sense, there is no question about this being a highly efficient, productive economic system. Indeed, by the decision of the Brothers, the size of the market for Community Playthings and the total production has been deliberately limited. Similarly, because caring for animals involves daily obligations which might excessively separate Brothers from the community, a decision has been made not to keep milking cows or goats, though it might well have been economically feasible and desirable.

Men and women are spiritually equal within the Bruderhof. In joining the community as a full member, each person must do so in his own right. The sexes differ primarily in that each has its own functions, proclivities, and obligations. The difference is seen as one in destiny, not in equality.

To become a full Brother in the Bruderhof is likely to take a long time, but there are no rules. A newcomer may remain in the community for many years until he becomes a novice, surrendering himself fully to the cause. Thereafter, it may again take years until both the individual and the community are prepared for the new Brother. It is like a wedding: both sides must be prepared for the union, as well as the Spirit which is above them all.

It is similar with two young lovers who wish to marry. Not only they, but the entire church-community must be prepared for the establishment

---

[17] Eberhard Arnold, *Why We Live in Community, op. cit.,* p. 11.

of the new union. Without the support and participation of the total Brotherhood, there can be no marriage.

> The first grade of unity is the complete unity of spirit between God and the Church. Everything must be sacrificed for this unity of spirit. The second grade of unity is a unity of souls amongst all in the community of believers. For this second grade, the third grade of family unity must be sacrificed whenever necessary. The third grade is by no means despised or humiliated by the two greater unities; rather, the destiny of the third grade is to make the first and second grades understandable to those who otherwise have no way to approach them.[18]

Families, in the traditional, monogamous Western sense, are important within the community. The marriage bond serves as a reaffirmation, on a lower level, of the faith and the greater unity of the total community. Acts of marital infidelity are unthinkable. The bond between a man and a woman is inseparable because it is a bond which includes not only the husband and wife, but the entire community and the Holy Spirit.

Living in the Bruderhof is largely living separate, in another world. Television, radio, newspapers and magazines are available, but only sparsely followed. Young children are especially protected from the external environment. They are kept from hearing the news or from watching television except on very special occasions. Members of the Bruderhof do not participate in politics. They are conscientious objectors to war. Only when external circumstances are forced upon them in such a manner that they must respond, will they engage in direct confrontation. They resisted the imposition of a Nazi teacher upon their school, they refuse to fight in wars, and they will, if called, go to court to protest the construction of a new highway across their property. But they will never initiate legal actions against another person.

The Bruderhof does not view itself as a new church or a sect. The community is not affiliated in any formal sense with any organization or group. The creation of a new church would constitute but another human institution, another stumbling block between man and his Creator. Eberhard Arnold argued that the world-church made precisely this error when, toward the end of the second century, it formalized ecclesiastic hierarchies and became a power in world history.[19] But, insist

---

[18] Eberhard Arnold, *Love and Marriage, op. cit.,* p. 52.
[19] Eberhard Arnold, *The Early Christians* (Rifton, N.Y.: Plough Publishing Co., 1970) p. 8. Originally published in 1926.

the Brothers, it is only in the sanctification of daily existence, in bringing the Holy Spirit into every crevice of human activity, that the Church—God's Church, not man's Church—will be established on earth. "We have only one weapon to fight the depravity of conditions today. This weapon of the Spirit is constructive work in the community of love."[20]

The involvement of the community in the affairs of the world, to protest the war, to struggle for civil and human rights, to reduce poverty and pain, is the one area in which there is expression of mixed feelings among the Brothers,. On the one hand is the awareness that man alone cannot change the world, on the other is the compulsion to do something. Thus, material help and support is offered to those who are in need, but such efforts, while important, do not receive first priority. The government and other worldly institutions, including charitable organizations, are viewed as a necessity in an unredeemed world. The Bruderhof awaits a total reorientation, a revolution that will usher in the Kingdom of God. It is to the anticipation of this end that the Society of Brothers is dedicated. It is a world in which love, not power, is supreme. Since all temporal organizations are thus inevitably engaged in a struggle for power and domination over those who oppose and who disagree with them, the Bruderhof, while constituting one of the sharpest criticisms of modern life, can only very partially be a participant in the political and social struggles of our times.

### The Real and the Ideal

It might be contended that what stands above is not a description of the Bruderhof but only an idealization thereof—the face, the proscenium, without a view into the wings and the ante-rooms. It is not that another visitor, a different observer would have arrived at a different picture, but rather that all visitors are equally limited in their ability to penetrate what Eberhard Arnold more than once referred to as "the mystery of community." Mysteries cannot be observed, they must be experienced, and those who experience them have a different view of their reality than those who simply observe.

For the concerned and interested observer questions remain to which there are no simply answers. It may be suggested, for example, that the revolutionary idealism which induced the founders to rebel against middle-class European life can not be replicated in the second generation. Whereas for the first generation this idealism constituted a revolt,

---

[20] Eberhard Arnold, *Why We Live in Community, op. cit.,* p. 5.

for the second generation it requires conformity to the ideals of the elders. This source of conflict between the generations is a common phenomenon in revolutionary movements, yet the Brothers insist that among them there is no generation gap.[21]

Similarly, one may raise questions about the issue of leadership. The ideals and structure of the community are such as to promote collective decision-making. However, an academic observer living for an extended period of time within the community might discover not only a leadership hierarchy but also differential privileges. Perhaps he would become privy to the existence of cliques and factions. But he would discover them primarily because he is looking for them. At the same time, he is likely to completely dismiss the experience of the Holy Spirit which his positivist training has taught him to define as unreal and meaningless, the figment of other men's fantasy.

There are two issues here: (1) can a community be adequately described in terms of the ideals and the ideology that it sets for itself? and (2) can a community adequately be described in terms of categories of thought (in this case, humanistic sociology) which it consciously and vigorously rejects because of its own alternative conception of man and of reality?.

Neither of these questions can be satisfactorily resolved here. The answer is closer to no, than yes, on both counts. But that is simply affirming the fact that human knowledge is always incomplete. The issues raised apply to all attempts of human beings to understand each other, they are dramatized in this case because the discussion is of a significantly different social order with an alternative system of values. In this case, as in all others, we can perceive only that which is visible. We are required to comprehend these perceptions in our minds and in our hearts with sensitivity. This requires a sincere effort to experience the insights of others, as well as a conscious awareness that the "objectivity" that any one of us alone brings to a situation is but another subjectivity.

## Analysis

The Bruderhof was selected for study as a case example of an Intentional Community, as defined by the conceptual model. Intentional Communities are characterized by high goal orientation on the commu-

---

[21] One of the former children of the Bruderhof who is now an adult and has left the community has focused precisely on this issue in his comments on Bruderhof education. See Philip Hazelton, "Trailing the Founders; On Being a Second Generation Bruder," *This Magazine* (Toronto) 4:2 and 4:3, Spring and Summer, 1970.

nity and the family level, and low goal orientation on the external level. Our first task is to check the Bruderhof for fit to this model.

More specifically, the model leads us to expect the following characteristics:

1. Contractual partnership between levels II and III.

2. Specific goals on levels II and III (in the partnership).

3. Formal leadership hierarchy on level II.

4. No staff-resident split.

5. The exercise of coercive, utilitarian, and normative power on level II.

6. Roles of the subsystems (families and individuals) defined by the community, i.e., by level II.

7. Low goal orientation on level I.

8. No exercise of coercive or utilitarian power on level I.

9. Link of passive, generalized cooperation between levels I and II, and between levels I and III.

10. Level I defined by level II.

11. Horizontal links on level III primarily by contract.

At the very beginning of our analysis we are confronted by the central question: Does the goal of unity, and serving as a witness to the Kingdom of Christ on earth, constitute a specific goal in the Parsonian sense? Whereas there can be little doubt that in this case, the family and the community level are in partnership working toward the attainment of this goal, the question remains: Is this a *specific* goal?

What is the product of this goal? Does this product serve as an input into another system? Is the product amenable to contract? Is it measurable? Is there a way of knowing when it has been accomplished, or when it is being approached?

The answers to these questions, to the extent that there are answers at all, lie primarily in the realm of metaphysics, a realm which, for the Brothers, is at least as real as life itself. The product is the visible witness to the Kingdom of God on earth. It serves as an input into another "system", God.[22] Its accomplishment is deemed possible only because Christ on the Cross has made a contract with mankind, sending His Holy

---

[22] Only secondarily do the Brothers make an input into society at large, serving as witnesses before their fellow men.

Spirit to the community of believers, in their unity.[23] The pursuit of unity is never-ending, but every Brother is intimately aware of when and under what circumstances greater unity can be achieved, and also when he and the entire community fail.

It is difficult to reconcile these goals with the Parsonian definition of specific goal. Indeed, Parsons might argue the exact opposite, insisting that the Bruderhof has no goal other than the satisfaction of the needs and preferences of its own members. Hillery thinks of intentional communities as having purposes, but not specific goals.[24] Similarly Kanter, in a somewhat different context, refers to this ideological aspect of 19th century utopian communities as a part of the communal, the *Gemeinschaft* dimension, not as a part of the associational, the *Gesellschaft* dimension. For Kanter, it is the associational dimension which transcends the boundaries of the community and assures its survival within its environment; it is the associational, e.g., the economic, and the politically reformist dimension, which is concerned with inputs and outputs, not the communal.[25] If, therefore, we persist in seeking to identify the goals of the Bruderhof as specific goals, we run counter to a major trend in sociological thought.

It may be best not to pursue this particular issue further, until the other points in the analysis of the Bruderhof have been touched upon.

The model leads us to expect a formal leadership hierarchy on level II. In the Bruderhof there is no such hierarchy. There are stable leadership roles, but there is also an ideologically based, absolute denial of hierarchy. In the Bruderhof, roles are to be equated with functions; status is not accompanied by privilege. It must be concluded, therefore, that in this case, the expectations with respect to hierarchy posed by the theoretical model are not met.

In the Bruderhof there is no hierarchy . . . but two thoughts make us hesitate. First, in reflecting, we find that in this respect the Bruderhof is unique among long-lived Intentional Communities. Most other commu-

---

[23] Christ taught, "believe ye in Me and live." Salvation is conditioned upon faith, according to this interpretation of the Gospel. In this sense there is a contract, an agreement to limit cooperation. Witness the fact that for centuries there has been a debate among theologians as to whether and to what degree "works" are included within the contract.

This is a Christian reformulation of the Old Testament idea of the Covenant. The Covenant is a treaty, a contract, between man and God: I will be your God and you will be my people; I will guard over you if you will serve only Me.

[24] George A. Hillery, Jr., *The Family and the Residential Community: The Case of the Commune,* (unpublished paper, 1970) p. 34.

[25] Rosabeth Kanter, *Utopia, A Study in Comparative Organization,* (doctoral dissertation, Department of Sociology, University of Michigan, 1967) p. 23f.

nities of this type have hierarchies, authority structures, and/or formal leadership roles with differential rights and privileges.[26] Secondly, one should be suspicious of the concerted persistence with which the presence of a leadership hierarchy is denied by the Brothers. It is not that one needs to suspect that there really exists a hierarchy, and that it is being kept secret, but rather, that the denial of hierarchy obviously is related to the specific ideology of the Bruderhof. The consistent demand for equality in rights and privileges among all the Brothers in this case makes the development of a formal hierarchy an impossibility. In other words, had a different Intentional Community been selected for analysis, it is probable that we would have discovered a more clearly defined, formal leadership hierarchy.

There is no staff-resident split within the Bruderhof; about that there can be no doubt. In this respect the Bruderhof meets the expectations of the model. Within the community there is no staff, there are no individuals representative of the external level, and therefore there is no possibility of a split.

In the analysis of social control mechanisms, we are confronted with another problem. The model leads us to expect the exercise of coercive, utilitarian, and normative power within the partnership system, i.e. on the community and the family level. But utilitarian power is non-existent within the Bruderhof, because there is no private property, no clear line of demarcation between "mine" and "ours". Because of its practice of pure communism, material rewards and punishments are not available either in principle or in practice. This leaves only normative and coercive sanctions which might be exercised within the community. The exercise of normative power within the Bruderhof is ubiquitous. But coercive power exists only in the sense that individuals or families can be forced to leave the community, i.e. excommunicated. Precisely because the community has no way of exercising utilitarian power, normative controls become essential. The problem is that the ultimate and only coercive sanction is too costly; in each case it can be used only once, and if it is over-used, it may threaten the life of the total community, as at Sannerz in 1922.

As anticipated, the communal organizational subsystems, i.e., the families, while largely conforming in their organizational structure to the

---

[26] Among Kanter's nine "successful," i.e. long-lived, utopian communities, four had explicit hierarchies. The other five "successful" communities each had a single, powerful leader. Kanter, Dissertation, *op. cit.,* p. 123-124. Also, Rosabeth M. Kanter, *Commitment and Community,* (Cambridge: Harvard, 1972) pp. 116-125.

patterns established by Western society, are conceived as being subservient to the total community. The writings of Eberhard Arnold make this perfectly clear. Again, the extent to which individuals and families live up to these ideals is not the central question. The fact is that the community reserves unto itself the right to control, and therefore limit, the organizational structure of its families, and its claim to "higher unity" is, in principle, agreed upon by the Brothers as an essential element of their ideology. In this case, the marriage contract is a genuine contract in that it constitutes an agreement not only between the marriage partners, but also between these two and the community. In this case the marriage contract limits the cooperation of the marriage partners in that their love and unity is of a lower order than the unity of the community.[27]

With the exception of the commercial ties of the toy factory, the major source of economic subsistence, the Bruderhof has few contractual linkages with the external world. In order that the demands which they place upon the community may not supersede the demands of faith, the goals of production and financial profit remain explicitly subservient to the theological goals of the community.[28]

Level I, as represented by external political and economic institutions, is defined by the community as a necessary, and therefore a tolerable phenomenon in an unredeemed world. Every effort is made to minimize the impact of external systems upon the life of the community and its members. Only in the area of high school and college education has the community permanently resigned itself to making compromises.

There is another, less explicit way in which the Bruderhof has chosen to limit its isolation from its environment. The move from Paraguay to North America was largely motivated by the desire to leave the culturally alien environment of rural Latin America for a more nearly receptive and compatible world.

---

[27] See Hillery, *Communal Organizations, op. cit.,* p. 57, where he discusses the marriage contract as a non-contract, because it characteristically imposes no limitations upon the relationship between the marriage partners. The link between the two is characterized by generalized cooperation because the family is a communal organization. In the case of the Bruderhof, however, the marriage contract is more nearly a genuine contract because it does impose limitations and the family is not fully a communal organization, in principle, not only in fact.

[28] By contrast, Oneida, one of the better known 19th century utopian communities, is today an industrial corporation. See, Maren Lockwood Carden, *Oneida: Utopian Community to Modern Corporation* (Baltimore: Johns Hopkins, 1969). Amana, another 19th century utopian community, had a similar history.

The Bruderhof has known times in which the benign, passive, low goal oriented, generalized cooperation of the external level was shattered. In Germany, under the Nazis, when level I became high goal oriented, it led to the expulsion of the community. Similarly, if in the 1960s conscientious objector status had not been available to draft age youth, the Bruderhof would probably have been forced to depart from this country. The Bruderhof can survive and thrive only as long as the external level (i.e. the community relevant elements of level I) is low goal oriented and non-coercive.

Since the family level is expected to be high goal oriented, we would expect contractual links among families to be predominant over links of generalized cooperation. The first and overwhelming impression given by the Bruderhof is that this is not the case. This is a community of love in which all of the interactions among families and individuals are permeated by sentiments of affection and acts of mutual aid. The model also leads us to expect contractual linkages between families and the community, since both levels are, in theory, high goal oriented and level III is expected to function as a formal subsystem of level II. Again, it is evident that families freely and voluntarily become members of the Bruderhof and that the linkage between them and the community may readily be viewed as one of generalized cooperation.

It would appear, therefore, that on these two counts also, the Bruderhof does not meet the expectations of the model. There is, however, another way in which these phenomena may be understood. This second way which is suggested by the model is admittedly hypothetical. It is based upon a different interpretation of the interactions on the family level which cannot be claimed to have been fully validated by these data. Since this study has focused primarily on the community level, analyses and hypotheses which relate primarily to the family level can only be projected with great caution. With these important qualifications in mind, the following alternative analysis of linkages in the Bruderhof is suggested. The alternative analysis is proposed not simply because it makes our case conform more closely to the theoretical model, but also because it appears to us to be a reasonable interpretation of the data. Whereas it may correspond less with how the members of the Bruderhof interpret their own actions, it is in our judgment a more adequate explanation of their actual behavior.

Our alternative analysis would contend that among the Brothers mutual aid constitutes a constant, never-ending, high goal oriented task.

Primarily through it, is the all-important *unity* achieved. It is unity which makes the potential presence of the Holy Spirit within the community a reality. Without unity there can be no holy witness.

We perceive in this instance the special case of a contract of *total* cooperation, which is not to be confused with active generalized cooperation. Total cooperation is contractually defined as cooperation having no limits; i.e., an agreement to cooperate in all things; generalized cooperation, by contrast, is limitless because it has not been defined. Within a contract of total cooperation we must observe a fine, but important distinction. There is no limit to the areas of cooperation but there *is* a limit to the agreement: it specifically prohibits non-cooperation. By contrast, within generalized cooperation the parties to the agreement are free to cooperate or not. Whereas a contract of total cooperation is characterized by mechanistic interaction, generalized cooperation is characterized by structural free wheeling.[29]

Our assertion that in the Bruderhof, not generalized cooperation, but a contract for total cooperation prevails rests upon the finding that *non-cooperation among members and families is explicitly proscribed.*

A contract is an agreement to limit cooperation. As such, it defines the areas of cooperation and, by indirection, the areas of non-cooperation. Whereas in most contracts, the area of cooperation is relatively small and the area of non-cooperation includes all other fields of interaction, in the case of a contract for total cooperation, as in the Bruderhof, the area of cooperation ideally includes *all* fields of interaction and the area of non-cooperation is defined, by indirection, as including nothing.

The vertical linkage between individuals and families on the one hand, and the community on the other, is of a similar nature. Initiation into the

---

[29] Hazelton, who grew up in the Bruderhof, in his analysis and criticism of his childhood experiences ditinguishes between friendship and community: "Kids believe in friendship, not community," he writes. "I've never seen a community of friends; but I've seen a community of enemies." Hazelton, *op. cit.,* p. 60. " ... friendship has to be one of the things a community seems to have to de-emphasize, if not actually combat." *Ibid.,* p. 67. What Hazelton calls "friendship" is a link of active generalized cooperation; "community" is a contract of total cooperation.

It is of interest to contrast, in passing, the idea of friendship as mentioned by Hazelton, and the idea of marriage among the Brothers, discussed earlier. Both marriages and friendships constitute a source of potential threat to unity and to the contract of total cooperation. But marriages are legitimated by the community and incorporated into the contract of total cooperation, constituting a "lower level of unity". On the other hand, friendships among youth, percisely because they are not, or cannot be incorporated into the contract of total cooperation, because they are exclusive links of generalized cooperation, serve as a threat to the unity of the whole.

Brotherhood is viewed as a process which is similar to that of a Bruderhof wedding.[30] The event celebrates an agreement, a contract for total cooperation between the individual and the community. This contract which seals the partnership between the family level and the community level is, again, intimately related to the goals of the community.

Let us now return to the question of whether there is a specific goal on the community level. Three possible approaches to this problem lie before us: (1) to reject the model, (2) to insist that serving as Holy Witness to the Kingdom of God is a specific goal, and thus change the definition of specific goal, or (3) to retain the model, indicating those areas where the case example does not fully fit, and suggesting both areas for possible emendation of the model, and areas for further empirical research. We will examine each of these options separately.

(1) It appears to be unwarranted to take the first option. It would be premature to reject the entire model at this point.

(2) If the ideological goal of the Bruderhof is considered to be a specific goal, then we must also insist that God is a social system which (Who) utilizes the output of the community as an input. We are on very shaky ground. Clearly, God, whatever He is, is not a social system. Whereas theologians have for ages discussed the question, does God need man, does He need man's prayer, his Holy Witnesses, etc., there is absolutely no basis for our entering into this discussion here.[31]

We conclude that it is inappropriate to subsume under the Parsonian meaning of specific goal abstract, ideological, metaphysical phenomena such as these. There would be little sense in attempting to revise Parson's definition. And finally, neither is there any value in discussing whether these phenomena, these metaphysical outputs of the community level, are inputs into level I (society), or into level I double prime (a metaphysical realm).

(3) The course that remains open to us is to re-examine the question of whether the absence of a specific goal in the Parsonian sense must necessarily place an organization into the category of communal organization. The question is raised with specific reference to an organization, such as the Bruderhof, which on the community level exhibits other

---

[30] i.e., not an ordinary wedding which leads to a relationship of generalized cooperation, but a Bruderhof wedding.

[31] This is not meant to totally exclude consideration of such phenomena by social scientists, especially to the degree that belief in their reality is found to influence human behavior.

characteristics which would lead us to classify it as a formal organization.

In his adaptation of Parsonian theory to the concept of communal organizations, Hillery focusses almost entirely on the nature of the specific goal and not upon the other distinguishing attributes of communal, as contrasted with formal organizations which are discussed in Chapter 2, above.[32] Moreover, Hillery focusses on the phrase, specific goal, to the virtual exclusion of the phrase, oriented towards. The latter, it is suggested, refers to the functional and structural characteristics of an organization, such as hierarchy, types of power used, and the nature of linkages, rather than to its output, the specific goal.

It may be reasonable to suggest, therefore, that an organization such as the Bruderhof is high goal oriented without, in fact, having a specific goal in the Parsonian sense. Even though the output is in this case not measurable, not empirically verifiable to outsiders, it is to the Brothers. They know when and under what circumstances they are oriented toward the goal and they know how well they succeed (or fail) in their efforts to achieve it. On the other hand, many outsiders are likely not to recognize an ideological goal as being empirically identifiable and measurable.

In order to distinguish it from a specific goal, let us call this an *ideological goal.* An ideological goal is the output of a system which is structured in the manner of a formal organization. The system gives primacy of orientation toward it. It is specific and identifiable, especially from the point of view of the members of the system. It serves as an input into something external to the system, but not necessarily into another social system.

In the case of the Bruderhof, serving as Holy Witness through unity constitutes an ideological goal. The community and its members in all their actions attempt to give primacy of orientation toward this goal which is from their viewpoint specific, identifiable, and measurable.

If, on the contrary, the goals of the community were different, if its goals were happiness, joy, goodness, self-expression or some other broad ideal, then these would not be considered ideological goals and the

---

[32] In his most recent writings Hillery has tacitly recognized the difficulty of focussing on the goal rather than on structure. His concern with the issue of freedom in communal as contrasted with formal organizations bears witness to this new direction. See, George A. Hillery, Jr., "Freedom and Social Organization: A Comparative Analysis," *American Sociological Review,* 36:51-65, 1971.

community would probably not be high goal oriented. These goals are not specific and they are not considered, under normal circumstances, to serve as inputs into an external system. The key question that would remain is, whether the hypothetical community oriented toward happiness is structured in the manner of a formal, rather than a communal organization.[33]

There is sense and value to introducing new concepts only to the degree that they more adequately give expression to empirical reality. Whereas the concept of an ideological goal appears to be useful to the analysis of the Bruderhof, it may not be similarly applicable to that large category of communities that is presumed to fit into the Intentional Community category, the 19th Century utopian communities. But the picture looks outwardly promising. It may be suggested, for example, that in Oneida originally the ideological goal was primary, and economic (specific) goals were secondary. Over time, as a consequence of *goal displacement,* the economic interests of the community and of individuals took precedence, and the ideological high goal orientation declined.[34] This would suggest that in Intentional Communities the communal aspect is given expression not in the ideology, i.e. in high goal orientation towards an ideological goal, but in the crevices between, in the areas not regulated by either economic necessity or overtly ideological constructs.

It further suggests that the mechanisms of commitment identified and classified by Kanter might be viewed as high goal oriented, institutionalized acts directed toward the attainment of the ideological goal of the community.[35] They are a part of the Gesellschaft rather than of the Gemeinschaft aspect of community. According to this view, the mechanisms of commitment are acts performed in fulfillment of the contractual partnership between the members and the community. The mechanisms of commitment are to generalized cooperation (Gemeinschaft) what a T-group is to intimacy—mechanisms, but not the genuine experience.

Let us recapitulate this analysis of the Bruderhof. Among the eleven criteria listed at the beginning of this section, the Bruderhof failed to

---

[33] An interesting case in point is Sun City, Arizona, a retirement community whose ideological goal might be considered to be happiness. See, C. Trillen, "Reporter at Large," *The New Yorker,* 40: 120-70, April 4, 1964.

[34] Goal displacement is here used in the traditional sense as it applies to formal organizations.

[35] Rosabeth Moss Kanter, "Commitment and Social Organization: A Study of Commitment Mechanisms in Utopian Communities", *American Sociological Review,* 33, August 1968, pp. 499-518.

meet three. The Bruderhof has no hierarchy on the community level, it has no specific goal in the Parsonian sense, and some major questions remain concerning the matter of linkage on the family level. The model does appear to fit the other eight criteria fairly well.

The obvious solution to the conflicting views of the Bruderhof which have been presented is to say that both positions are partially valid. It is a mixed case. On the community and the family level the Bruderhof may be said to partake of some of the characteristics of both formal organizations and communal organizations. This is a solution, but it avoids the major issue.

The key question is, does the behavior of members of the Bruderhof more nearly resemble that of elements of an inclusive level formal organization, or that of members of a communal organization? In our view, the question is not so much how the members feel or think about their role, as how they behave. In this connection, the key questions which remain to be more fully tested are: (1) to what degree is non-cooperation among the members proscribed? (2) to what degree is all behavior related to the ideological goal of the community? (3) what, if anything, constitutes the area of autonomy for families, beyond the right to excel in cooperation or to dissociate themselves totally from the community?

We have been led to favor the view that the Bruderhof can best be understood as a social system which is high goal oriented on the family and the community level. We have reached this tentative conclusion on the basis of the finding that non-cooperation among members appears to be proscribed, that all behavior appears to be maximally related to the ideological goal, and that the autonomy of families is significantly circumscribed. But before our conclusions can be stated more definitively, more research is required, not only with the Bruderhof, but also in other, similar communities.

There is an additional reason for the tentativeness of our conclusion. The Bruderhof may not have been an appropriate case for the testing of our model. In taking another look at the Bruderhof's partnership between the family level and the community level, we find that it constitutes a rather unequal, a lopsided contract. The importance of the community level is much greater than that of the family level. This suggests that it might be better to consider the Bruderhof a borderline case, between the Intentional and the Total Community. The Total Community, it will be recalled by the reader, is an anti-community in which the community level is high goal oriented (in this case toward the ideological goal) and the other two levels are excluded. In the total community there is no partnership, no contract between the levels.

It must now be noted that, in an important respect, the Bruderhof is not similar to the 19th century American utopian communities. Whereas the goals of most of the utopian communities included some form of individual salvation, or self-fulfillment, the Bruderhof specifically denies this point. To the Bruderhof salvation (a word that the Brothers try to avoid) is potentially achieved only in unity, in and for the community as a whole.

In the Bruderhof there are nearly no goals at all on the family level, separate and distinct from the ideological goal of the community. In the 19th century utopian communities, by contrast, there are both solidary and reciprocal goals within the contractural partnership between the two levels. Individuals and families joined these communities giving up former possessions, associations, and comforts in exchange for the promise of salvation which, they were led to believe, was made available to them only via the community. The Bruderhof is different in that the contract for total cooperation has the effect of merging level III into level II until, ideally, the two are fully united into one. Is this not what Eberhard Arnold meant when he said that community is a mystery, a *unia mystica* in the medieval sense?

There is insufficient evidence, but a case might be made for the view that the Bruderhof began as an Intentional Community and moved at an early stage, probably after the crisis of 1922, in the direction of a Total Community. The movement was only partial, as is evidenced by the fact that the families and individuals are not totally excluded, not totally dependent upon the decision of the community. The movement in the direction of a Total Community, the partial process of decommuniza- tion, may explain why the Bruderhof, unlike most utopian communities, has remained nearly intact for nearly half a century. This view would lend modest support to our research hypothesis, that deviant communi- ties are unstable within their category. The Bruderhof started out as a mixed type, a deviant community, but moved in the direction of one of the anti-communities, thus gaining greater stability. One of the major differences between the Bruderhof and the 19th Century utopian com- munities appears to lie in the fact that the latter did not experience the same kind of decommunization.

The utopian communities did not form a part of this study and therefore it would be unwarranted to state a finding or conclusion which includes them. It might be suggested, however, that one of the reasons the utopian communities in most instances did not survive over long periods of time is related to the fact that they had reciprocal as well as solidary goals between the community level and the family level. Over time, the reciprocal goals became dominant. By contrast, in the Bruderhof there is only a single, solidary goal which rests primarily with the community level.

The utopian communities, according to this view, are more like associations, exhibiting the structural dilemmas discussed by Olsen.[36] Individuals and families are brought together into the community because they share common purposes. However, the community, like an association, can never satisfactorily serve the interests of all the individuals. Over time, collective interests develop which differ in part, from those of the individual members, thus creating tension between the family level and the community level. The Bruderhof (except during its first two years) is different, because there the idea of unity, of collective action on the community level, precedes in time, in concept, and in priority the interests of individuals. It cannot be stressed too much that the Bruderhof demands not simply self-denial, but more important, the affirmation of unity. Those who do not identify with this goal either do not join, or do not remain within the community.

---

[36] Mancur Olsen, *The Logic of Collective Action* (New York: Schocken, 1968).

# CHAPTER 9

## SUMMING UP

Again and again as one comes across studies of communities one is reminded of their delightful and endless variety. Every community has a unique personality of its own: empirical reality is always much more colorful and more complex than theory.

In telling the stories of Levittown, Moosehaven, and the Bruderhof, the objective was two-fold: to check each case against the theoretical model for fit, and to pilot test the community change hypothesis which was proposed in Chapter 5. Since both these issues have been discussed in detail in connection with each of the case studies separately, in this summary chapter we are left with a discussion of the implications of these findings for the model as a whole. The broader implications for community theory and for planning will be discussed in later chapters.

### Testing for Fit

Both Moosehaven and Levittown, i.e., the Administered and the Designed Community, appeared to fit the anticipated model without difficulty. It turned out that the Bruderhof, the Intentional Community, was, from a theoretical standpoint, the most interesting. The Bruderhof study led to a suggestion for revising the original approach to testing the distinction between formal and communal organizations. The work of Hillery had encouraged us to examine the *output* of the organization, the specific goal as defined by Parsons. Our revised view changes neither the definition of specific goal, nor the fundamental conceptual distinction provided by Hillery. Rather, it changes the focus of the analysis from the output as a measurable something, to the *orientation* of the system under study.

The changed focus does not alter the criterion of primacy of orientation toward a specific goal, but it does alter the significance of the often difficult and confusing discussion of what, in fact, is the specific goal. In Moosehaven, is the specific goal building the Order, or caring for aged and dependent members? What is the specific goal of a local government, or of a school? According to this new focus, the significance of these questions is reduced and analytical emphasis is placed instead upon an examination of the functional and structural characteristics of the organization such as linkage, hierarchy, and the types of power legitimately used.

The concept of ideological goal was introduced to help explain structures and behaviors within the Bruderhof which led it to resemble more nearly a formal than a communal organization. An ideological goal is not a specific goal in the Parsonian sense because it does not lead to the production of a measurable output into another social system. However, in the eyes of the members of the system, it is specific, measurable and identifiable.

It remains to be seen how useful the concept of ideological goal, as defined here, is in the analysis of other Intentional Communities. Since the Bruderhof was found to be a borderline case, approaching the Total Community, it may not have been an adequate test of Intentional Communities in general.

The concept of contract for total cooperation (an agreement to cooperate in *all* things) was introduced to provide an alternative explanation of horizontal relationships on the family level in the Bruderhof. The validity of its application to this case rests primarily upon the finding that non-cooperation is maximally proscribed.

In order to more fully test the concept of contract for total cooperation in the Bruderhof, as well as in other Intentional Communities, additional measures will need to be developed on the family level. Such measures must, in part, be capable of assessing deviance from institutionalized norms and accepted patterns of cooperation within the community. Under a contract for total cooperation, in theory, the only deviation from norms allowable is that of surpassing others in their cooperation. Those who choose to deviate in other ways are sooner or later excluded from the community because, in effect, they are acting to exclude themselves.

Whereas the model fit surprisingly well in the case of Levittown and Moosehaven, it should be tested further, with additional case examples.

Prisons, 19th century company towns, Oak Ridge, Tenn., British New Towns, and in several years, Columbia, Md., and Reston, Va., might be studied in detail as other examples of Administered and Designed Communities. Many of the newer cities in developing areas of the world would make especially interesting case studies, e.g., Brazilia, Islamabad, Nairobi, and Tashkent. In these cases it is not always readily apparent whether they more nearly resemble Administered or Designed Communities.

Additional empirical examples of Intentional Communities, as defined by the model, might clarify some of the unresolved issues raised by the case study of the Bruderhof. Are all Intentional Communities oriented toward an ideological goal? How common are contracts for total cooperation in such communities? Is it characteristic of other contemporary Intentional Communities to deny the existence of hierarchy on ideological grounds?

A study of one or more 19th century utopian communities in terms of the definitions which have been developed here would shed important additional light upon the model. Similarly, other contemporary religious communites such as the Menonites and monasteries need to be examined. The difficulty with including contemporary urban and rural communes is related, in important part, to their relative youth.

## The Historical Development of Deviant Communities

The hypothesis led us to anticipate that deviant communities would change over time, either in the direction of Crescive Communities, or in the direction of one of the anti-communities. It now appears that each of the three cases is different, and few general conclusions can be reached concerning the issue of historical development.

Moosehaven seems to have shifted from the Intentional to the Administered category, exhibiting a slight tendency toward Totalitarianism. The Bruderhof shows signs of a partial movement in the direction of the Total Community. Levittown experienced a change in the nature of its external level. There remains some ambiguity as to how Levittown should be classified today, though its tendency appears to be in the direction of the Solipsistic Community.

The hypothesis was based upon the assumption that deviant communities are unstable in their category. It was anticipated that either the partnership would deteriorate as a result of one of the two levels taking control (decommunization), or the partnership would erode, leading toward the Crescive Community. The latter process, it was hypothesized, would result from the product of two processes, goal erosion on the community level, and democratization, i.e. the elimination of the staff-resident split.

It cannot be claimed that the hypothesis was validated. On the other hand, it also has not been refuted. None of the three communities is today as it started out in the beginning. The Bruderhof and Levittown give evidence of partial decommunization. Moosehaven has moved laterally, from one deviant category into another, accompanied by a minor tendency toward decommunization. None of the communities give evidence of having moved in the direction of the Crescive Community.

These deviant communities are not as unstable within their categories as had been anticipated. It is true that each of the communities experienced a major shift. Curiously, each of these shifts took place at a time rather early in the community's history. Since that time, relatively little systemic change has taken place. It is almost as if each of the three communities had to find a particular "niche". Perhaps it is precisely those communities which do not find their niche, that do not long survive. But unfortunately such a conclusion does not lead any closer to being able to specify a particular community's niche at the outset.

### Some General Observations

1. This has been a study of communities. Both the methodology and the model utilized in this study can not be assumed to have similar applicability to all other levels of social system analysis. It was noted in the discussion of Levittown that a determination whether the external system, as represented by the totality of American society, is high or low goal oriented cannot be made in the terms suggested here. It is a question too highly interwoven with political and philosophical value judgements.[1] Similarly, but for different reasons, the measurement of high or low goal orientation on the family level is fraught with difficulties. On the other hand, there is reasonable expectation that this analytical approach may be of value in the study not only of communities, but also of other types of organizations on the mezzo level, i.e., organizations located between the total society and the family. It might, for example, be applied to the analysis of voluntary associations, or ethnic groups.

---

[1] However, see Chapter 12, in which such judgements are attempted.

2. Level I, the external level, is always important in its broader dimension. This is merely another way of saying that no organization, no community, in the long run escapes the hand of history. Every one of the communities studied changed significantly because of changes in the larger society. For example, the residents of Moosehaven, on the average, became more dependent. The Bruderhof experienced several moves, and Levittown could not have been created prior to 1947. Whereas in this study, level I has generally been identified as that system or complex of systems external to the community which most directly affects its being and survival, the external level, in the larger sense, always includes all of American society, if not the entire world.

3. The word deviant as applied to communities has been used throughout this study, perhaps without adequate justification. Usually this term is applied to the behavior of individuals whose actions do not conform to socially accepted norms. In utilizing the word deviant to apply to a social system, i.e., a community, this study has introduced a new, a special meaning. It is important to be explicit.

A social system is deviant if it is structured in a manner which is contrary to societally accepted norms for that type of system. Thus, a corporation is deviant if it is structured like a communal organization; a family is deviant if it is structured bureaucratically, like a corporation. The use of the term deviant community is consistent with this definition, except as it applies to anti-communities. Theoretically, the anti-communities are also deviant communities, but in order to emphasize their greater degree of deviance, we have chosen to call them by a separate term.

4. This study has focused on the family as the primary subsystem of the community. The decision to concentrate on the family was based upon Hillery's definition of the vill as a "localized system integrated by means of families and cooperation."[2] Other subsystems of the community such as clubs, commercial institutions, ethnic groups, and churches were not made an important subject of this analysis. A short word about them is in order.

Community subsystems, other than the family, were found to be of significance only in Levittown. In both the Bruderhof and Moosehaven they are little in evidence. If we divide these community subsystems into the two types, formal and communal, we find almost no formal

---

[2] Hillery, *Communal Organizations, op. cit.,* p. 65.

subsystems—and few communal subsystems—within the Bruderhof and Moosehaven. In Levittown, on the other hand, both formal and communal subsystems are strongly in evidence: churches, lodges, clubs, commercial organizations, governmental subunits, etc., as well as families and friendship or ethnic groups. Nearly all of the formal and communal subsystems, but especially the formal subsystems of Levittown, are extensions of external systems which reach into the locality. Strictly speaking, they are not subsystems of the community at all; they are subsystems of external systems, located in the community. What Warren called the "Great Change" has determined the nature of Designed Communities most directly, Administered Communities indirectly, and Intentional Communities least of all.

# CHAPTER 10

# THE IDEA OF COMMUNITY

Modern America is probably one of the worst places at one of the worst times in history in which to attempt the study of communities. The many popular and technical meanings of the word community are so confused and confusing that it is tempting to avoid using it altogether. But this is perhaps the least important of the reasons that the discussion of community has been made difficult.

### The Community Movement

More to the point is that the word community has, especially during the past decade, acquired a value-laden meaning: specifically, the meaning that was excluded from the discussion in Chapter 3 above, "community as sentiment." Philip Slater, for example, speaks of the "desire for community," which he identifies with "the wish to live in trust and fraternal cooperation with one's fellows."[1] For Robert Nisbet the word community functions as an antonym for personal alienation.[2] In the daily speech of an ever increasing number of individuals the word community is to be understood as meaning love, togetherness, sharing, frankness and truth. In these and similar uses of the word, the reference is not to a type of human group, but rather to a desired and valued end.

Analytically, it is necessary to distinguish between two aspects of community as sentiment, the vertical and the horizontal. Vertically, it is identical with what has earlier been called sentimental collectivity orientation. It is a variable, equally applicable to both formal and

---

[1] Philip E. Slater, *The Pursuit of Loneliness* (Boston: Beacon Press, 1970), p. 5.
[2] Robert A. Nisbet, *Community and Power* (New York: Oxford University Press, 1953, 1962). See especially, the Preface to the 1962 edition.

## PLATE 9

### *COMMUNITY AS SENTIMENT IN COMMUNAL AND FORMAL ORGANIZATIONS*

Community as Sentiment

|  | horizontal aspect | vertical aspect |
|---|---|---|
| Communal Organizations | active generalized cooperation<br>e.g., mutual aid; love; brotherhood<br><br>A | sentimental collectivity orientation<br>e.g., commitment to the group<br><br>C |
| Formal Organizations | (*)<br><br>e.g. sentiment among assembly line workers<br>B | sentimental collectivity orientation<br>e.g., patriotism; company spirit<br>D |

* In most formal organizations the horizontal aspect of community as sentiment is relatively unimportant. See the discussion of the human relations approach, and of Homans' in Appendix A.

communal organizations. A commercial organization, a factory, a military unit, a family or a hippie commune might all be characterized by a high (or low) degree of loyalty, allegiance or commitment to the inclusive system (i.e., vertical community sentiment). On the other hand, only in communal organizations is the horizontal aspect of community as sentiment of primary importance. Horizontally, community as sentiment may be equated with active generalized cooperation. (See Plate 9).

Unfortunately, horizontal/vertical distinctions are not regularly made by the advocates of the community movement.[3] Vertical community as sentiment enhances what Landecker called "normative integration", i.e., agreement between the standards of the group and the behavior of its members.[4] When taken to extremes, normative integration is potentially oppressive and pernicious, as in the case of ultra-nationalism.

---

[3] Community movement is here defined as consisting of those persons who identify with the concept of community as sentiment, who view community as a valued and desired objective.

[4] Werner S. Landecker, "Types of Integration and their Measurement", *American Journal of Sociology*, 56:4 January 1951, pp. 332-40.

Horizontal community as sentiment is active generalized cooperation on level III, i.e., among families. It promotes what Landecker defined as "communicative integration," i.e., fluidity in the exchange of meanings among the members of the group.[5]

To the extent that the aim of the community movement is to overcome personal alienation and to build trust and fraternal cooperation, horizontal community as sentiment appears to be of far greater significance than vertical. Practically, the efforts of the community movement need to be directed toward strengthening active generalized cooperation within communal organizations (Box A, Plate 9), and to attempt to reintroduce or strengthen horizontal community as sentiment within formal organizations (Box B).The strengthening of vertical community as sentiment (Box C) serves the community movement's objective only to the degree that it supports the horizontal, and does not displace it. The strengthening of sentimental collectivity orientation in formal organizations (Box D) seems to be irrelevant, if not contrary to the aims of the community movement.

It will be recalled that the total absence of active generalized cooperation on level III is the common characteristic of all anti-communities. In the anti-communities, horizontal community sentiment cannot exist. Therefore, the major objective of the community movement might be stated in social-structural terms as the desire to prevent, retard, or undo the perceived contemporary trend toward the anti-communities, i.e., to prevent decommunization.

### The Small Town as Crescive Community

There is another reason why the study of communities is particularly difficult in modern America. It seems that whenever one uses the word community, the much romanticized image of the American small town comes to mind. It is probable that such idealized communities never existed in the form in which they are remembered. For example, the small town in upstate New York studied by Vidich and Bensman, seems to be remarkably unattractive.[6] But more to the point is the fact that

---

[5] Landecker has identified two other types of integration: "functional integration," i.e., reciprocality in the exchange of services among members of a group, and "cultural integration," i.e., consistency among the cultural standards of a group. Functional integration is enhanced by the proliferation of contracts. Speculatively it might be suggested that cultural integration is enhanced by affirmations of ideology.
[6] Arthur J. Vidich and Joseph Bensman, *Small Town in Mass Society; Class Power and Religion in a Rural Community* (Princeton: Princeton University Press, 1958).

such small towns, to the extent that they continue to exist, have become largely irrelevant to this nation of cities and suburbs. In short, had the attempt been made to include a small town as an example of a Crescive Community in this study, it would probably have been as atypical of American life as were the three deviant communities.

Perhaps one day the small town will again assume an important place in America, but for the present, the Crescive Community might better be sought elsewhere. In the Age of Retrenchment it is likely that large parts of our urban population will be redistributed across the land, but for the moment it may be more important to examine the "neighborhood movement" as the possible locus of new Crescive Communities. The new communal structures that have grown out of the Community Action Programs, the Model Cities Programs, and the recent efforts to decentralize metropolitan school systems bear careful study.

Whereas neighborhoods might be regaining significance among low income families, the non-geographic community has retained or attained (it is not clear which) great importance in middle class circles, e.g., the Jewish Community, the intelligentsia, all city planners as professional peers, or all Spanish speaking persons within a predominantly English speaking municipality. It might be highly appropriate to seek the contemporary Crescive Community in this important, non-geographic dimension of American society.

### Community as Place and Space

Hillery, in his discussion of the vill, identified three foci: family, cooperation and space. This study has largely followed Hillery's advice in concentrating on the family and on cooperation. However the concept of space has received somewhat less attention. Space in vills suggests a physical area with boundaries. Each of the three cases studied occupies such an area, though it is not always clearly defined, as in Levittown, or is subject to change, as in the case of the Bruderhof. Space is important, and communities such as the Gypsies and Medieval Jews should also be considered deviant communities, quite independent of the definition of deviant community which has been developed here. They are deviant communities because for them space constitutes more nearly an interchangeable and non-continuous area, than a specific physical expanse with boundaries.

It would appear that space, as one of the three foci, is not as significant for purposes of structural analysis as the other two. This does not, however, make it unnecessary or unimportant. By means of this focus the community, a social phenomenon, is related to the land, a physical phenomenon. In its primary concern with human interactions, sociology has often overlooked the inanimate, the physical world. But to pursue this point further here would take us too far astray.[7] Let us add only that, at what level, by whom, and according to what principles is space allocated, are always among the key questions to be raised in the analysis of a community. Perhaps this is because space is both essential for survival, and within the community (and in the world) always limited. Unlike food or shelter, the supply of space is usually not humanly expandable, except by means of external conquest.

### Community and Freedom

Freedom is another of those words which in everyday language has acquired such a multiplicity of meanings that it is difficult to use it without inviting confusion. Most broadly, freedom is universally understood to represent a good and desirable human condition or objective. It has psychological as well as sociological overtones. Thus, one person may feel free in a given place, and under given conditions, in which another person does not. Freedom is the recognition of necessity, proposed the philosopher Spinoza. Viewed differently, freedom is increased as one's options are increased.

These and other meanings of the word freedom might, not without difficulty, be reconciled. But this is a task better left to others. In speaking of community and communal organizations the idea of freedom was introduced as a part of Hillery's concept of structural free wheeling. Structural free wheeling has been projected as a sociological, rather than a psychological concept. When more fully explored, it might serve a similar function in sociology to that of Heisenberg's principle of indeterminacy in physics. It is important because it is related to spontaneity and creativity, aspects of being human which are among the most highly valued in our society.

In the past, the dominant impulse among social scientists has been to impute causal relationships between phenomena, much in the manner of the early mechanistic ideal of the physical sciences. The recognition that the appearance of randomness in human interaction is not due to the

---

[7] See, Alvin W. Gouldner, "The Theoretical Requirements of the Applied Social Sciences," *American Sociological Review,* 22:1, 1957, pp. 92-102.

inadequacy of scientific knowledge or to the inherent complexity of the phenomena, but rather, that randomness is a basic principle inherent in certain types of social structures, is one of the most important philosophic assumptions underlying this study. It declares that human beings and the institutions which they create act and interact only partially in mechanistic response to each other. Descartes' glorious vision of a clocklike universe is confronted and limited by an alternative vision of man, a being in the universe who *must* create, and therefore is unpredictable.[8]

Structural free wheeling is not equated with freedom; it is merely viewed as one dimension thereof. It follows that in deviant communities there is somewhat less freedom in the sense of structural free wheeling than in Crescive Communities. In anti-communities there is the least freedom. Quite possibly, the members of some deviant communities *feel* freer than they would if they lived elsewhere. This is unquestionably the case in the Bruderhof, but it is unrelated to the concept of freedom as structural free wheeling. This special meaning of freedom is emphasized because it is importantly related to the discussion of the limits of planning to which we are now prepared to return.

---

[8] This view has been given expression most persuasively by the philosopher Susanne Langer. It is perhaps no accident that the first truly prominent woman in the entire history of Western philosophy (Xanthippe notwithstanding) should be the source of these ideas. *Vide,* Susanne Langer, *op. cit.*

# CHAPTER 11

## THE LIMITS OF PLANNING

To plan is to exercise control. To plan beyond the limits of planning was defined in the introductory chapter as the attempt to exercise control in areas which, for reasons that extend beyond technical or political limitations, *should not* be planned. Planners were called upon to become moral philosophers, to make moral judgments as to when, and under what circumstances, they are overstepping the limits of planning.

Within the near future, an Age of Retrenchment is envisioned in which both the technical capacity for and the need for more control through planning will be greatly enlarged. In the past it was usually taken for granted that planning is better than non-planning. But in the new era the question, when and under what circumstances does planning threaten freedom, emerges with renewed significance.[1]

To suggest a simple dichotomy between planning and freedom is both false and misleading. Planning may serve either to restrict or to expand freedom, depending on the circumstances, on the role of the planners, and on the definition of freedom.

Planning by experts on the inclusive level, i.e., central planning, serves to expand freedom to the extent that there is consensus within the system.[2] Thus, for example, highways and traffic laws, though they impose controls upon all drivers, have the aggregate effect of expanding freedom. There is consensus among all drivers (and most, but not all non-drivers) that highway and traffic planning are good and desirable.

---

[1] For a classic discussion of this problem see, Karl Mannheim, *Man and Society in an Age of Recontruction* (New York: Harcourt, Brace & World, 1941) pp 364-381.

[2] See Barbara Wooton, *Freedom Under Planning* (London: George Allen and Unwin, 1945) for a detailed discussion of this issue as it applies to central planning in England.

The opportunity to expand freedom, defined as the multiplication of options, under central planning is determined by the existence of consensus of two types. There must be general agreement on the rightness of the specific action, e.g., on the need and desirability of traffic controls and on what constitutes reasonable control. In addition, there must be general agreement on the method of establishing and enforcing controls, e.g., on the authority of traffic engineers and policemen, on the legitimacy of the courts, etc.

But the degree of consensus among Americans which, until recently, has been maintained around the issue of highways and traffic, is exceptional. In our society it is more common for individuals and groups to be in contest with each other because they perceive their interests to be in competition. Within this context, planners have in many cases taken an advocate stance, they have become political actors promoting the objectives of specific constituencies.[3] Advocate planners aim to expand the power of their constituents, thus to enhance their freedom by multiplying their options. An economic scarcity model of power (and of freedom) is implicit in this conception. All power seekers are assumed to be in competition with all others: what one gains the other loses.

Within advocacy planning consensus is assumed on the inclusive level: all actors are expected to agree to the rules of the contest which are imposed by broad societal norms.[4] But dissensus can and often does arise among the actors in their competition for scarce, but commonly valued, objectives. The freedom of the elements on the included level is circumscribed both by the amount of power which they possess in relation to each other, and by the degree of consensus on the inclusive level.

The limits of planning, the areas which should not be planned because they threaten freedom, are defined for the central planner as those areas which lack broad consensus. For the advocate planner these limits, it might now be suggested, are defined by the degree to which he/she contributes to the more in equal distribution of power within society.

---

[3] It is often argued that *all* planners are advocates of specific interests, that the idea that planners can be objective, unbiased experts is untrue and misleading. See, Paul Davidoff, "Advocacy and Pluralism in Planning," *Journal of the American Institute of Planners*, November, 1965. By contrast, Altshuler argues for the expert role. Alan A. Altshuler, *The City Planning Process* (Ithaca: Cornell, 1965).

[4] What John Friedmann called "counterplanning" is advocacy planning within which there is disagreement on the rules of the contest. John Friedmann, "Notes On Societal Action", *Journal of the American Institute of Planners*, 36:5, September 1969, pp 311-318.

That is, advocate planners who serve only the interests of those who possess high concentrations of power are acting beyond the limits of planning.

But the concern for freedom which was expressed in the first chapter remains only partially resolved. Undoubtedly, in the Age of Retrenchment greater consensus within broad issue areas such as conflict resolution, human welfare, and the use of non-renewable natural resources, will be essential to survival itself. The equal redistribution of resources and power must become a primary, principled goal of all planning. In short, there can be no meaningful freedom without a broad level of consensus on order, and there can be no true freedom without equality. But even as the attempt is made to satisfy these basic needs, to guarantee these basic freedoms, planners must remain sensitively aware of the human need for freedom as defined by structural free wheeling, the opportunity and the capacity to act spontaneously and independently without a necessary concern for reaction.

We might reemphasize at this point that freedom as structural free wheeling is not only an individual, subjective phenomenon. Rather, it is to be understood as a type of linkage between and among the elements of a social system. These elements may be individuals (as in a microsystem), or they may be groups (as in a macrosystem). Structural free wheeling is to be social-structurally conceived: it is maximized among the elements of communal organizations.

It follows that communal organizations are beyond the limits of planning. Only formal organizations can be planned. Structural free wheeling cannot be programmed. But again, the world is always more complex than theory. It is not neatly divided into formal and communal organizations; planning invariably affects the life and death, the growth or decline of communal organizations. Though communal organizations, in theory, cannot be created because they must generate themselves, often they are in fact created and planners do step beyond the limits of planning. The result appears to be the planning of deviant communal organizations.

In this connection we return to the discussion of the planning of communities, specifically new communities.[5] They effectively illustrate

---

[5] The phrase new community is here meant to include New Towns, communes, "open" institutions, and many other contemporary efforts to resolve social problems by means of new, total environmental designs.

the dilemmas that planners face who may initially be interested only in planning cities, housing developments, or towns, but whose actions invariably have the latent effect of creating communities.

### The Planning of Communities

Inevitably the planner of a new community will begin with a deviant community. This is primarily because planners function within formal organizations which have, at least initially, a specific goal: the creation of a new community. Crescive communities cannot be planned, and the anti-communities are not desirable and therefore there is no reason to plan them. The planner's choices are therefore restricted to one of the three deviant types.

But there are times when new communities, i.e., deviant communities, should be planned. Especially if it can be shown that the new community will satisfy important needs and desires of the intended residents, then its planning can be justified.

In thinking about the residents of deviant communities it appears that each community type might be specially suited for a specific type of population. The Administered Community is designed to accommodate a dependent population, and a passive, alienated personality type. The Intentional Community accommodates an ideological population, an aggressive, socially committed, ideational personality type. The Designed Community accommodates an aggressive, individualistic population, persons who tend to be sensate and materialistic.[6]

The above social-psychological typology requires further evaluation and empirical testing, of course. But, it suggests the idea that in the future, different types of communities might be planned specifically to serve different types of populations. Depending not only on personality type, but also on age, familial circumstances, health, religion or ideology, individuals may choose to live in different types of planned communities. Whether this represents a new and unwanted form of social segregation or the way to greater human fulfillment is a question that must, of course, be carefully considered.

---

[6] Pitrim Sorokin's three basic cultural types are related to this classification. Sorokin's sensate, ideational and idealistic cultures may be related to the Designed, the Intentional and the Crescive Community, in that order. Not unlike many historians of the first half of the twentieth century, Sorokin overlooked the fourth type, the Victim culture, which is pervasive in history and is to be paralleled by the Administered Community. See, Pitrim A. Sorokin, *Social and Cultural Dynamics* (New York: American Book Co., 1937).

Thus, deviant communities might be thought of as having special functions in soceity. Presumably in most instances, the long term goal of new community planners is not to create a deviant community. In these cases, paradoxically, the planners' aims are to maximize their control with the objective of ultimately minimizing it. Only in this way might they enable a Crescive Community to generate itself.

### The Columbia Experience

Columbia, Maryland has begun to serve as a model for a multiplicity of American new town experiments. Therefore, it seems appropriate to take a brief look at the Columbia experience in the light of the analytical system that has been developed in this study. We are greatly assisted in such an analysis by the work of Richard Brooks, who was at one time the social planning consultant to the Rouse Company, the developer of Columbia.[7] Interestingly, Brooks emphasizes precisely those points which are of special significance for our analysis.

Columbia began with what Brooks calls a "total institutional phase." As Brooks defines it, the characteristics of this phase resemble closely what this study has called an Administered Community. Columbia differs from Levittown in that Rouse was interested, among other things, in the creation of a "good community"—not simply a good house at a good price.

Since the intended and actual residents of Columbia are not persons generally considered dependent, it is not at all surprising that the Administered Community phase did not last. Columbia, after it had absorbed its first residents, and with the inclusion and active involvement of a multiplicity of institutions which are not under the direct control of Rouse, entered a new phase.

Brooks distinguishes between, "those residents who see themselves primarily as consumers of a developer packaged community," as against those who believe that they have "purchased the right to participate in the decision of the new town." The first group appears to be pulling Columbia in the direction of the Designed Community. The second group is represented in part by young Antioch-Columbia students and faculty. Whether the latter group's purpose is to achieve something which more nearly approachesa Crescive Community, or something that approaches an Intentional Community, is not quite clear. The two

---

[7] Richard O. Brooks, "Social Planning in Columbia", *Journal of the American Institute of Planners,* XXXVII: 6, 1971, pp. 373-379.

positions might be stated succinctly in that the first group favors goal erosion on the community level whereas the second group emphasizes democratization. Brooks indicates that he believes that suburbia, i.e., goal erosion, will ultimately win out.

If Columbia goes the way of Levittown, Rouse and his planners are only partially responsible. The pull of the larger society is so strong that to counteract the trend would have demanded Herculean effort. Above all, Rouse had to survive financially and all other considerations had to take a back seat. Perhaps a time will come in America when financial considerations will no longer play such a key role; under those circumstances planners might be able to emphasize other goals than rapid return on investment.

In Israel, for example, the primary goal of the new towns has been the acculturation of new immigrants. Financial considerations played a secondary, or tertiary role.[8] Israel's new towns tend toward the Administered Community type; they primarily serve a relatively dependent population. In the long run, they will presumably become Crescive Communities.

The author has been involved in a unique American attempt to create a new town beginning as an Intentional Community.[9] The prospective residents are rural Blacks and much effort has been expended in attempting to gain their participation from the earliest planning stages of the undertaking. In this case also, the long range goal of most of the planners is not the maintenance of an Intentional Community, but democratization.

### The Future of Planning

Planners are persons who define problems and elaborate solutions in terms that they consider rational and feasible. Most familiarly they function within formal hierarchies and create new hierarchies. Planners are different from architects and engineers who design and create inanimate objects. Invariably, their work leads them to the creation of new, human organizations peopled by men. women, families, neighborhoods, communities—constantly demanding recognition and reaffirmation of their humanity and their right to freedom.

---

[8] Shimon S. Gottschalk, "Citizen Participation in the Development of New Towns: A Cross-National View," *op. cit.*
[9] Shimon S. Gottschalk, and Robert Swann, "Planning a Rural New Town in Southwest Georgia," *Arete,* 1:2, January 1971.

Planning is both inevitable and impossible. Without planning, there is little chance that our crowded, materially overexploited planet and its human population will long survive. With planning, it appears inevitable that planners will repeatedly be forced to cross the limits, and thus to threaten freedom.

Planners must become sensitive to the fact that danger threatens not only in the lack of adequate planning or in their inability to implement "good" plans, but also in overplanning. Constantly planners must remain alert to the need to leave room, as it were, in order that the communal organizations may generate themselves. As often as not, their purpose must be to gain control only in order to be able to give it away, or to lose it. This is a long step beyond "maximum feasible participation." At other times—and these may be the most important times—it means the dismantling of existing organizations in order that the functions and the solutions they have defined may be re-examined and reformulated at the roots.

The work of Ivan Illich who polemicizes against compulsory schooling for children effectively illustrates this need.[10] Illich distinguishes between schooling and learning, and asserts that the schools have monopolized the definition of learning, thus restricting, narrowing, and in his judgment deforming it. The work of the "true" educator, says Illich, does not lie in strengthening and expanding existing schools, but in the enhancement of a multitude of alternative opportunities for learning, and first and foremost, in the elimination of compulsory school attendance for children. The alternative opportunities for learning include not merely more diverse and more numerous independent formal organizations, such as private schools. Rather, what is more important, they require the encouragement of learning within the numerous, informal communal settings which abound, and which deserve support, rather than criticism, from those who profess to be educators.

Richard Sennet writes of the desire for "the freedom to accept and live in disorder."[11] It is similar to freedom as structural free wheeling. For Susanne Langer it is the recognition that man is a creature who *must* create. For Abraham Maslow the word is self-actualization.[12] For planners it leads to a paradox: the need to "rationally" plan for disorder; to "feasibly" unplan order; to rediscover, both within themselves and within all mankind, the joy of unpredictability.

---

[10] Ivan Illich, *Deschooling Society* (New York: Harper & Row, 1970).
[11] Richard Sennett, *The Uses of Disorder* (New York: Random House, 1970) p. XVIII.
[12] Abraham H. Maslow, *Toward a Psychology of Being* (New York: Van Nostrand, 1962).

CHAPTER 12

# TOWARD A SCENARIO FOR AN ALTERNATIVE SOCIETY

This is a postscript. The community studies, the analyses, the summary of findings and conclusions which stand above are not altered by this chapter. The purpose of this chapter is to present an act of reflection which, it is hoped, will shed light on some of the political implications of the ideas which have been discussed. If, in part, this chapter constitutes an exercise of poetic license, then let it be excused on the grounds that we are here dealing with the anticipated systematic elements of an as-yet-non-existent, alternative social order.

## An Alternative Normative Emphasis

The distinctions between formal and communal organizations which have been drawn in this study are philosophically related to a much larger tradition of dualistic thinking. It is the ancient contest between Hellas and Jerusalem, between man's attempt to subdue nature and his awe of it: secular—sacred, external—internal, adaptation—integration, instrumental—expressive, yang—ying, male—female. These approximately parallel dualisms, though drawn from a breadth of cultural sources and ideological traditions, express a common theme which is deeply rooted in the intellectual history of man.[1] The dualistic conception usually implies that the first member of the dyad is inextricably intertwined and in contest with the other. The two poles compete, and in competition support each other. They are both opposed and continuous.

It is the basic contention of the theoretical discipline which has become known as the sociology of knowledge, that man's reality is

[1] See also, Roland Warren, *Truth, Love and Social Change* (Chicago: Rand McNally, 1971) pp. 273-299.

socially constructed. Here again we find a dualism, though in this case of a somewhat different order. "In the dialetic between nature and the socially constructed world, a human organism itself is transformed. In this same dialetic man produces reality and thereby produces himself."[2]

Conceptions of reality become, over time, integrated self-perpetuating constructs surviving (and expiring) in history, somewhat like great empires. Berger and Luckmann demonstrate how, by means of the process of legitimization, an existing institutional order, an inherited conception of social reality, is granted "normative dignity."[3] The *status quo* tends to label itself "good" and to reject, subvert, redefine or coopt alternatives. Because we, as planners, thinkers and activists, are both subject and object within contemporary society, we must be especially conscious of the intellectual bias which our perspective imposes. As intellectuals in this society we almost inevitably contribute to the legitimization of the *status quo,* a process that has, "a normative as well as a cognitive element."[4]

Let us now bring together these two themes, philosophic dualism and sociology of knowledge. Here we join a long line of philosophers and historians in the contention that the history of cultures, epochs and civilizations can be viewed and interpreted in the light of the dualistic conception. Historical periods differ from each other in that one or the other pole achieves normative emphasis or dignity. The legitimacy of the medieval world, for example, was based upon a complex of theological justifications of an institutional order which may be identified with the communal, expressive, internal pole. With the Reformation, the rise of mercantilism and industrial capitalism, Western society began to move toward the opposite pole. In contemporary society formal organizations, instrumental relationships, and adaptive thinking have achieved normative emphasis. They are justified by means of a legitimating symbolic universe which values, among other things, success, progress, power, change (within legitimated bounds), and material possessions.

If the above assumptions and theoretical postulates are accepted, then the theory of communal organizations is likely to be useful in developing an alternative conception of reality. It is a way in which we may attempt to view the seeds of the future and the past, in the present. The currently normatively emphasized, self-perpetuating formal organizations exercise

---

[2] Peter L. Berger and Thomas Luckman, *The Social Construction of Reality* (New York: Doubleday, 1966) p. 183.
[3] *Ibid.,* p. 93.
[4] *Ibid.*

all forms of power (normative, utilitarian, and coercive) to retain control. The competing, currently normatively de-emphasized communal organizations can, according to our theory, undermine established power by challenging its legitimacy only by exercising normative power, e.g., through acts of non-cooperation. If they attempt to exercise other than normative power, then they will begin to resemble formal organizations, and their efforts to gain normative emphasis will have been subverted.

We need to distinguish clearly between high and low goal orientation on the one hand, and normative emphasis and de-emphasis on the other. High and low goal orientation constitutes a basis for the systematic distinction between formal and communal organizations. By contrast, normative emphasis relates to questions of societal norms, of alternative value systems. Either high or low goal oriented systems may be normatively emphasized in a specific society, i.e., either formal or communal organizations may be normatively more valued and cognitively more legitimate.

Throughout the foregoing chapters of this study, it has been assumed that the high goal oriented levels possess and retain normative emphasis. There was little reason to assume otherwise. This has been a study of American communities which form a legitimated part of contemporary American society. Though this has been a study of deviant communities, none of these granted normative emphasis to low goal orientation.[5]

In the present chapter, alternative normative assumptions will be made. A symbolic universe consisting of alternative thought structures (the cognitive element) and alternative values (the normative element) will be conceptualized. Following the model, we will sketch out a scenario for an alternative society and of community types within it in which the low goal oriented levels and the communal organizations are cognitively more legitimate and receive normative emphasis.

The importance of attempting to envision such an alternative society has been given impetus by the recent work of Roland Warren in his analysis and discussion of the likelihood of innovative change in the Model Cities program. Warren suggests two "Diagnostic Paradigms." Paradigm I represents the current symbolic reality and Paradigm II its alternative. Warren concludes that programs such as Model Cities:

---

[5] A different way to make this point is to indicate that all three of the communities, each in a different way, subscribed to what is generally referred to as the Protestant ethnic: it is good to work, to suffer, to strive for the goal.

are bound to have extremely little effectiveness in changing social conditions so long as they do not help to create alternative institutionalized
thought structures based on different diagnostic paradigms and containing
components which are internally as supportive of the alternative paradigms
as are the components of Paradigm I.[6]

We agree with Warren's analysis of the problem, but not entirely with
his approach to a solution. We agree that from the perspective of
Paradigm I (contemporary society) it is not possible to develop Paradigm
II (alternative society) solutions. But we go further by asserting that it is
impossible to *create* alternative institutionalized thought structures such
as Warren is seeking, in the sense that planners create. Warren suggests
that the alternative knowledge base to the present "system paradigm" is
to be provided by an "actor paradigm." This far we agree. Warren's
actor paradigm would, "emphasize the competitive struggle of individual
actors . . . for their own well being."[7] Here we disagree. According to our
conception the actor paradigm would emphasize generalized cooperation
among communal organizations, not competition among individuals.
The components supportive of the alternative institutionalized thought
structures are the communal organizations. They cannot and they need
not be created. They are already among us, but they need to achieve
normative emphasis and cognitive legitimacy. In others words, the
alternative society, like the communal organizations discussed in earlier
chapters, is beyond the limits of planning.

## Conceptualizing an Alternative Social Order

Contemporary American society is probably one of the most highly
goal-oriented societies that has ever existed. We are high-goal oriented,
and our culture defines this as something desirable, and good. That is, in
America high goal orientation is normatively emphasized. We are proud
of being the most powerful and the richest nation on earth. We "believe"
in ambition, in the unequal social value of human beings, and we readily
justify the use of coercive power under the persuasive slogans of "law
and order" and "keeping the peace." Social phenomena such as these are
generally viewed not as necessary evils, but as positive, desirable
"goods."

---

[6] Roland Warren, The Sociology of Knowledge and the Problems of Inner Cities, *Social Science Quarterly*, 52:3, December 1971, pp. 469-91.
[7] *Ibid.*, p. 491.

Because our value system favors high over low goal orientation, our cognitive structures, even our language, exhibit this bias. The nation is commonly equated with all those who are encompassed by the state. The community is often equated with local government, and in analysis, primarily the formal, rather than the communal structures are taken into account. Ours is a society in which the corporation, that most ingenious of all legal inventions, has permeated nearly all areas of life, including at times the family. The corporation, a paper "person," is often considered more real than a living individual. It has immortality, people do not. An individual's identity is largely derived from his linkage to particular formal organizations; a person without papers is a non-person. In America, when the question is raised, how much is so-and-so worth, the expected response is related to the size of his financial holdings, expressed in measurable, contractually exchangeable, monetary terms.

These observations are not new, nor different from those that have been made by others. Our purpose is not to improve, or to expand upon them. They are repeated only to remind the reader of the heavy value bias which permeates our society (as much as any other conceivable society).

Let us attempt to imagine an alternative society which, it must be admitted at the outset, is somewhat like having an intellectual dream. It is important to underline again that the differences between high and low goal orientation, between normative emphasis and normative de-emphasis, between contemporary society and its alternative are consistently to be considered differences of degree. As in other dyadic social relationships, the two poles co-exist, both contesting and supporting each other. Just as the present society does not represent a social order in which high goal oriented systems are *always* normatively emphasized, nor does the alternative society represent a social order in which the opposite is *always* true. Empirically it has been shown that high goal oriented systems consistently include and require the presence of low goal oriented systems. Similarly, low goal oriented systems, if for no other reason than the need for physical survival, require the high goal oriented systems.

In embarking upon our reflective exercise, to imagine the alternative society, we are immediately confronted by the recognition that a minimal degree of physical security must be presupposed. But this appears to be a reasonable assumption, applicable to the Age of

Retrenchment in which production will be planfully curbed, and presumably more equitably distributed.[8]

The alternative society is one in which the low goal oriented systems, the communal organizations, are normatively more emphasized (valued, dignified) and cognitively more legitimate. Therefore, the accumulation of resources for the exercise of coercive or utilitarian power is negatively valued. Equality rather than hierarchical differences is emphasized. Style rather than outcome, process rather than goal, are applauded. Contracts are less "binding" than bonds of generalized cooperation. The institutions of such a society would be accorded respect and honor to the degree that they are created and constantly recreated by those whom they serve. Institutions which for reasons of efficiency or expediency are imposed by, or inherited from others, would be accorded the least normative dignity.

It is more difficult to anticipate the alternative cognitive structures which would accompany the alternative society. The word "we" might acquire as many synonyms as the word "money" in our present society. New, non-task-oriented roles for individuals and social systems would proliferate. Alternative ways to achieve a goal would be of greater interest than goal achievement itself. To succeed in the contemporary sense of the word, would be to fail, because it would imply forsaking one's brothers, elevating oneself above them, exercising other than normative power over them, in short, denying being for the sake of doing.

The above sketch of the alternative culture is based upon an attempt to logically invert the presently legitimated normative and cognitive structures. The result is especially interesting because it reveals an alternative cultural outlook which in current empirical reality is represented especially by those who might be called victims in the existing social order: rural blacks, poor Appalachian whites, Chicanos, and the peasants of the underdeveloped nations—numerically, perhaps the majority of mankind. These people, if they accept the normative and cognitive structures of contemporary Western society, are doomed to

---

[8] In speculating about the alternative society in this, and the following pages, analogies and examples will be drawn from a variety of counter-cultural settings. These observations have not been verified. This identifies a task that lies before us at some time in the future: an analytical study of counter-cultural institutions.

Contemporary counter-cultural communities presuppose material sufficiency, and perhaps security as well. On this point see, Philip Slater, *The Pursuit of Loneliness,* pp. 103-4.

almost certain failure. On the other hand, if they accept and internalize the alternative societal values, then they will begin to succeed in their own terms, building a new world in which *their* values will achieve normative emphasis.[9]

Let us try to express this differently. The oppressed will fail as long as their aim is simply to replace the oppressors in their positions of

**PLATE 10**

*THE TRANSLATION OF COMMUNITY AND ANTI-COMMUNITY*
*TYPES FROM CONTEMPORARY TO ALTERNATIVE SOCIETY*

| CONTEMPORARY SOCIETY | | | | | ALTERNATIVE SOCIETY | | | | |
|---|---|---|---|---|---|---|---|---|---|
| CLASSIFICATION | NORMATIVE EMPHASIS(*) | | | TYPE | CLASSIFICATION | NORMATIVE EMPHASIS (*) | | | TYPE |
| Administered Community | I | high* | P | Deviant Community | Jet Set Community | I | excluded | | Anti-Community |
| | II | high* | | | | II | excluded | | |
| | III | low | | | | III | low* | | |
| Designed Community | I | high* | | Deviant Community | Traditional Village | I | excluded | | Anti-Community |
| | II | low | P | | | II | low* | | |
| | III | high* | | | | III | excluded | | |
| Intentional Community | I | low | P | Deviant Community | Voluntary Slave Community | I | low* | | Anti-Community |
| | II | high* | | | | II | excluded | | |
| | III | high* | | | | III | excluded | | |
| Crescive Community | I | low | | Theoret. Norm | Anarchist Community | I | low* | | Theoret. Norm |
| | II | low | | | | II | low* | B | |
| | III | low | | | | III | low* | | |
| Orwell 1984 | I | high* | | Anti-Community | Paradise | I | high | | Deviant Community |
| | II | excluded | | | | II | low* | B | |
| | III | excluded | | | | III | low* | | |
| Solipsistic Community | I | excluded | | Anti-Community | Artists Commune | I | low* | B | Deviant Community |
| | II | excluded | | | | II | low* | | |
| | III | high* | | | | III | high | | |
| Total Community | I | excluded | | Anti-Community | Vagabond Community | I | low* | | Deviant Community |
| | II | high* | | | | II | high | B | |
| | III | excluded | | | | III | low* | | |
| Totalitarianism | I | high* | | Anti-Community | Totalitarianism | I | high | | Anti-Community |
| | II | high* | | | | II | high | | |
| | III | high* | | | | III | high | | |

KEY:
- I = External Level    P = Partnership
- II = Community Level    B = Brotherhood
- III = Family Level    high = high goal oriented
-                        low = low goal oriented

[9] These lines should be understood in the manner of an hypothesis to be verified. Simply by making these assertions we lay no claim to having proved our case.

The alternative value system of the poor has been conceptualized as follows: "To put it simply, the lower class individual doesn't want as much success, knows he couldn't get it even if he wanted to, and doesn't want what might help him to get success." Herbert Hyman, "The Value Systems of Different Classes", in *Class Status and Power,* Reinhard Bendix and Seymour Lipset, eds., (New York: Free Press, 1966).

authority and power, i.e., if they try to "succeed." Even if they unite and win the struggle, they will have lost because they will merely have replaced the actors, and not changed the plot; the result will be simply a new class of oppressors and oppressed. The alternative society, by contrast, calls for the kind of transvaluation of values which is discussed above.[10] In such a society, the oppressors will have no foothold because they do not receive tacit approval and adulation from those who are oppressed.

The analysis which has been begun in this chapter cannot be finished here. The purpose has been simply to indicate the direction of the alternative thought structures and value systems, without entering into details. Indeed, perhaps too much has already been said because, as we insisted earlier, the alternative society, in principle, cannot be planned, least of all by those who have a vested interest in the present.

But since this is a study focusing on communities, one more step is indicated. We must try to sketch, even if only briefly, the alternative types of community structures which are to be anticipated in an alternative society.

### A Classification of Communities in the Alternative Society

In contemporary society the high goal oriented levels of the community are consistently expected to be normatively emphasized. The question raised in this subsection is, what types of communities and anti-communities are expectable in a society in which the high goal oriented levels are not similarly emphasized. What happens when institutionalized normative and cognitive thought structures are primarily supportive of the low goal oriented levels?[11]

In the alternative society, the classification of community types which was developed in Chapter 4, above, is completely turned upon its head. The former deviant communities become more like anti-communities. Three of the former anti-communities become deviant communities, and only the Crescive Community and Totalitarianism remain within their type. As in Chapter 4, we are here discussing ideal types.

In the alternative society there are no longer any partnerships between high goal oriented levels. Instead, "brotherhoods" between two or more

---

[10] The transvaluation of all values called for by Nietzsche *(Umwertung aller Werte)* has long ago taken place: we worship the powerful, despise the weak, and blame the victim. This is a new call for the transvaluation of all values, reversing and rejecting Nietzsche.

[11] In the discussion which follows it will be helpful to the reader to follow Table 2.

low goal oriented levels are in evidence. A brotherhood might be understood as consisting of two or more levels united into a single domain of generalized cooperation. Brotherhoods are always normatively emphasized. The levels which formerly constituted the partnerships are now excluded.

Before we proceed with a brief discussion of each of the alternative community and anti-community types it should be noted that we are here engaged in a mental exercise which has its conceptual parallel in the work of futurologists such as Herman Kahn and his colleagues at the Hudson Institute.[12] The method is similar, although, because we begin with different assumptions, the outcome is very different. Following a few logical rules, we take our previously defined variables, transpose them, and examine the outcome. The exercise is worth the effort because the outcome *is* interesting. As the names of the alternative society community types are intended to indicate, they are to a limited degree empirically verifiable as counter-cultural types in the far corners of the existing society.

The second justification for pursuing this exercise is to give greater substance to the otherwise vague notion of the alternative society. Of special importance is the discovery that the alternative society does not constitute an idyllic utopian dream. It may be preferable to the present, but it remains in large measure a highly problematic, complex social order.

The three alternative anti-communities which evolve from the inversion of the former deviant communities differ only modestly from their analogous contemporary anti-community types. The major difference is that the low goal oriented level attains and retains its position of normative emphasis without the exercise of coercive or utilitarian power. These are anti-communities because they embody the total denial of high goal oriented systems—a new form of oppression available only in the alternative society. They are unrealistic because, in purely material terms, their chances of survival are minimal.

The *Jet Set Community* is much like the Solipsistic Community. Though their life style is normatively emphasized and dignified, families in this community are unconsciously dependent upon others. The picture is much like that of a community of the super-rich in our own day. The *Traditional Village* is similar to the Total Community. In the contempo-

[12] Herman Kahn and Anthony J. Wiener, *op. cit.*

rary world it is represented by the tradition-bound communities which are normatively resisting the incursions of modernization in the developing countries. The *Voluntary Slave Community* places low value upon itself and upon its members, and worships and serves the external system. Again, the similarity to Orwell 1984 is apparent. It is Erich Fromm's *Escape from Freedom* institutionally incarnate.[13]

The former anti-communities are of greater interest because in the alternative society they become the deviant communities. In *Paradise,* the family and the community level form a brotherhood. It is like some of the contemporary retreatist communities located in New Mexico or Vermont.[14] In the *Artists Commune* the external and the community level are low goal oriented and normatively emphasized. Great value is placed upon universal esthetic "truth" and upon living the "beautiful life" in the brotherhood of all mankind. Individualism and family exclusiveness are negatively valued. The *Vagabond Community* is much like the present-day wanderers who populate the nation's highways. Among them, personal bonds and unstructu1 ed universal ideals are normatively emphasized. Local societies with their high goal oriented systems are deemphasized.

*Totalitarianism* remains an anti-community in the alternative society, but it is normatively deemphasized and therefore powerless and irrelevant beyond its own boundaries.

Of special interest is the *Anarchist Community* which is the former Crescive Community. Here, in one brotherhood, as it were, all three levels are united within a single low goal oriented, normatively emphasized system. In contemporary society the Crescive Community serves as a romantic memory of what never really existed: the low goal oriented, small-town American dream is largely irrelevant to present reality. By contrast, in the alternative society the Anarchist Community becomes the cultural ideal.

### A Reassessment

The attempt to systematically define the nature of the alternative society which appears above must be understood primarily as an agenda for future research, not as a set of logically or empirically derived conclusions.

---

[13] Erich Fromm, *Escape from Freedom* (New York: Reinhard, 1941).
[14] It might be suggested by some readers that the Bruderhof belongs in this category. We would disagree. The Bruderhof belongs where is has been classified, as in Chapter 8, above. But *perhaps* the Monastery does belong here.

To the extent that the taxonomy of alternative society communities is illustrated by the elements of the contemporary counter-culture, it has been influenced especially by the writings of Philip Slater, Theodore Roszak, and Charles Reich.[15] These authors are united in the view that the counter-culture constitutes a revolutionary force, and an important challenge to existing cultural norms.[16]

The idea that the value system of the alternative society finds expression among the victims of the existing social order is likely to strike the reader as an item of faith, rather than a verifiable truth. It is based, in part, upon the writings of Oscar Lewis, who was the first to introduce the concept of the culture of poverty.[17] Unfortunately Lewis' idea of the culture of poverty has been used during the past decade primarily by apologists for the *status quo*. Rather than blaming systemic imbalances and injustices for the persistence of poverty in America, they have blamed the victims. Programmatically, they have sought to impose changes upon poor people, rather than upon the social and economic system which is the source of, and which perpetuates, their poverty. We wish to totally dissociate ourselves from this view and from this political position.[18]

But because a theory has been used for politically undesirable ends does not invalidate it as a theory. The phrase culture of poverty is here used not as an explanation for poverty, but only descriptively. Whereas poverty in a world where there is potential plenty is an evil, the defined culture of those who suffer in poverty need not be similarly negatively viewed. The conservative apologists view themselves and their own culture as superior (legitimate) and the culture of poverty as inferior (illegitimate). But in the alternative society, the culture of poverty is normatively valued and emphasized, and the culture of the present *status quo* is normatively deemphasized.

All broad generalizations such as the ones which have speculatively been suggested here are dangerous, because they overlook important

---

[15] Philip Stater, *op. cit.;* Theodore Roszak, *The making of a Counter-Culture* (New York: Doubleday, 1969); Charles Reich *The Greening of America* (New York: Random House, 1970).

[16] This view has, of course, been challenged, most especially by "establishment liberals" like Daniel Bell, and Bruno Bettleheim. See Kenneth Kenniston, "A Second Look at the Uncommitted," *Social Policy* 2:2, pp. 6-19, 1971.

[17] Oscar Lewis, *Five Families* (New York: Basic Books, 1959), p. 16f. and *The Children of Sanchez* (New York: Random House, 1961) XXIV.

[18] These views are examined in detail by Stephen M. Rose, *The Betrayal of the Poor* (Cambridge: Schenkman, 1971).

aspects of empirical reality. To speak of the culture of poverty or of peasant culture in general terms is to ignore the fact that such "pure" cultures do not exist. Peasant societies are always parts of larger, more complex cultural wholes. Moreover, they can generally be characterized as containing dual traditions, an elite culture which is often related to physically and historically distant elements, and a lay culture.[19]

Granted the observation that societies and social systems are always more complex than they appear on the surface, let us suggest the following simplified, schematic presentation of societal types as they relate to the single variable of normative emphasis. (See Plate 11).

## PLATE 11

### *SOCIETAL TYPES AND THEIR NORMATIVE EMPHASIS*

| TYPE OF SOCIETY | NORMATIVE EMPHASIS |
|---|---|
| 1. Primitives and peasants below the caloric minimum. | 1. High goal orientation |
| 2. Traditional folk village | 2. Low goal orientation |
| 3. Modernized folk village | 3. Mixed |
| 4. Urbanized villagers | 4. Mixed |
| 5. Middle class suburb | 5. High goal orientation |
| 6. Counter-cultural commune | 6. Low goal orientation |

The concept of caloric minimum was suggested by Eric Wolf.[20] This minimum is estimated to consist of 2,000 to 3,000 food calories daily. It is not met by a large proportion of mankind. Families and societies existing below the caloric minimum are assumed to place normative emphasis upon high goal orientation because theirs is a constant struggle for physical survival. The southern Italian village studied by Banfield is a case in point.[21] Most primitive, nonagricultural societies presumably fall into this category.

The traditional folk village is the familiar subject of anthropological investigation. In such villages tradition, and the sacredness of heritage

[19] Robert Redfield, *Peasant Society and Culture* (Chicago: University of Chicago Press, 1960).
[20] Eric R. Wolf, *Peasants* (Englewood Cliffs: Prentice Hall, 1966) pp. 2-6.
[21] Edward Banfield, *The Moral Basis of a Backward Society, op. cit.*

and of land, are normatively emphasized. The modernized folk villages are these same traditional villages, but after they have experienced the impact of modern Western society. The urbanized villagers form ethnic communities such as may be found in the slums and ghettoes of large cities in most parts of the world.

In discussing the mixed cultures, especially the urbanized villagers, the familiar sociological distinction between caste and class may be helpful in explaining our view. A caste of victims, because its status is *ascribed* by unalterable heriditary factors is, according to our hypothesis, more likely to exhibit the values of the alternative society. Among a class of victims, on the other hand, status is ostensibly *achieved* (by their own malfeasance). Such persons are more likely to have internalized the value system of the currently legitimated social order. Most writings about poverty in America are concerned with this second phenomenon, not the first. In our view, among American urban blacks there is despair, *because* there is hope. But the hope is usually vain, and therefore the despair is self-destructive.[22]

Frantz Fanon comes close to a similar analysis of the problem of the underclasses. "Let us not pay tribute to Europe by creating states, institutions, and societies which draw their inspiration from her," writes Fanon.[23] But his approach to a solution undermines his analysis. When he declares that the starving *fellah,* the lowliest serf among the Arabs, *is* truth, then we appear to agree.[24] However, when he calls upon this same man to exercise violence as a "cleansing force," then he is in danger of undermining the very purpose of the revolutionary struggle.[25] By promoting the use of coercive power to achieve his ends, Fanon is in danger of laying the foundations of a new, potentially coercive, hierarchical social order, not an alternative society in which low goal orientation is normatively emphasized. In such a society it is unlikely that the *fellah* will *be* truth; rather, once again, he will be the victim.

But Fanon and the many others who align themselves similarly cannot be dismissed that easily. The anger and the frustration of the millions of

[22] Greer and Cobbs, two black psychiatrists, arrive at nearly the same view, but their solution partially differs from ours: William H. Grier and Price M. Cobbs, *Black Rage* (New York: Basic Books, 1968). Oscar Lewis based his conclusions on observations made in Mexico, not the U.S. Greer and Cobbs are analyzing the American urban black experience.

[23] Frantz Fanon, *The Wretched of the Earth* (New York: Grove Press, 1968) p. 314.

[24] *Ibid.,* p. 49.

[25] *Ibid.,* p. 94.

victims, both of the social system and of their own vain hopes, is *real.* It is not for us to instruct them how to act. Perhaps they must first pass through the disappointment that either victory *or* defeat is likely to bring them, before they can return to the rediscovery of the old, the alternative societal values. On the other hand, perhaps they are right. Against all odds, they may discover a way in which a society in which low goal orientation is normatively emphasized can emerge from a revolutionary struggle which is high goal oriented, and which exercises coercive power.

### A Dilemma and a Qualification

In the Introduction it was indicated that this study is addressed to both social planners and nonviolent activists. The opportunity has arrived to bring these two seemingly disparate interests together.

Both planners and activists are confronted by the same theoretical dilemma. Planners cannot plan communal organizations because communal organizations must generate themselves. Similarly, revolutionaries cannot *make* revolution, because making the revolution serves to undermine its outcome.[26] Planners commonly face the dilemma of either serving the *status quo,* or doing nothing. Revolutionaries face the dilemma of either *making* revolution and creating contradictions, or *living* the revolution and accomplishing next to nothing.

The moral and the practical implications of this dilemma are such as to humble all those who are conscientiously concerned for the creation of a more just and a more peaceful social order. There is no answer, except that each person and each group must discover its own place within the spectrum of positions that are possible, between being and doing.

In addition to the dilemma, we must introduce a qualification. It has been said earlier, that the alternative society does not constitute an idyllic utopian dream. Whereas the present society promotes inequality, coercion, competitiveness, and impersonal relationships, the alternative society promotes their opposites; yet the alternative society, too, has its price. It is likely to be characterized by reduced external incentives to action, inflexibility, caste rather than class divisions, and a concomitant degree of provincialism. There are times when the use of normative power in such a society is likely to be a suffocating experience for some

---

[26] The word revolution is here used to refer to the inauguration of the alternative society, as defined above.

of its members. In the alternative society, as in the present society, the moral-philosophic problem of defining the boundary between formal and communal organizations will remain.

If, in sum, the alternative society is preferred, then it is not because it provides solutions to all of mankinds's present problems, but rather because both the advantages *and* the liabilities that it portends at this point in history appear to be preferable to the inequities of the present.

## *A Note to the Non-Violent Movement*

There will be disagreement, of course, but a case can be made for the position that normative power is to be equated with the power of non-violence, that the communal organizations are non-violent institutions, and that the alternative society defines the goal of the non-violent revolution.

The problem lies, in part, with the words violence, and non-violence. They appear to have as many confusing meanings as the word freedom, which was discussed earlier. The phrase non-violence, which is derived from an unfortunate translation from the Sanskrit word *satyagraha,* does not, in theory, mean simply the absence of violence. Conversely, it leads only to confusion to refer to formal organizations as violent organizations.

*Satyagraha* is more properly translated as Truth Force, Love Force, or Soul Force.[27] Direct confrontation, non-cooperation, personal witness, symbolic protest, the arousal of public conscience are the primary means whereby the non-violent struggle has taken place. These are all methods of normative power.

Violence, though most commonly associated with the utilization of physical force, need not be conceptually restricted in this manner. Violence occurs when, by physical or symbolic means, a person's alternatives are willfully limited to a specific one. By this definition, violence usually coincides with the concept of coercion as used throughout this study. Formal organizations are not necessarily, or even usually, violent (coercive); it is only that violence is potentially, and in our society legitimately, available to them. This is not true of communal organizations.

A similarly clear-cut distinction cannot be made in the case of utilitarian power. This is expectable, since the differences between the

---

[27] Mahatma Gandhi, *Non-Violent Resistance* (New York: Schocken, 1961).

types of power are differences of degree, and utilitarian power occupies the middle ground. In theory, though probably not always in practice, those who profess non-violence would tend to deny the normative legitimacy of utilitarian power to almost the same degree as they deny that of coercive power.[28]

Until recently, the discussion of non-violent revolution has concentrated primarily on issues of conflict resolution, rather than on what Gandhi called the constructive program. But today, the term alternative institutions has come increasingly into focus, despite the fact that the theoretical nature of these alternative institutions remains largely undefined. This is where the theory of communal organizations may make a contribution. It suggests that an institution serves as a non-violent alternative to the degree that it grants normative emphasis to its low goal oriented elements, to the degree that it is non-hierarchical, exercises only normative power, is characterized by structural free wheeling, maximizes generalized cooperation, is defined from the bottom up (or from inside out), and encourages non-violent alternative institutions to flourish within it.

According to this view, the aim and the method of the non-violent revolution is toward and through the communal organizations. The end result will be not perfection, but a social order with new but preferable problems. It is not that the present society is totally a violent society and its alternative fully non-violent. Rather, the present society is one that legitimizes violence whereas its alternative legitimizes Truth Force.

From a political action standpoint, the program for the non-violent revolution begins not with the overthrow of the central government, but rather with the reaffirmation and strengthening of communal organizations at the roots, accompanied by normative resistance to coercion emanating from above. It means both helping to make it possible for others to build the alternative reality, and doing it oneself. This, for example, is where Women's Liberation plays an important role. Families cannot fully achieve the communal organizational ideal until women are fully equal. The same should be said, of course, of children, and of the aged. The revolution begins with the family, and spreads from there to friends and associates, to the community. Once it has taken hold there,

---

[28] This is an area of controversy. Witness the seemingly never-ending discussion among professed adherents of non-violence surrounding the question, does the destruction of physical property, such as draft records, constitute an act of non-violence.

then perhaps it will lead, by these entirely non-Marxian means, to the withering away of the state.

### Returning to the Beginning

This study began with a statement in anticipation of a new era, The Age of Retrenchment. As the political economic, and technological contradictions of our times come to a head, the resacrilization of our lives becomes increasingly essential to our survival.[29] Our Western world, which has become almost completely demysticized, may be on the verge of the rediscovery of its archetypal roots in the sacredness of the most simple and mundane human experiences. Howard Becker named such an alternative social order, the "folk-sacred society."[30] More appropriate might be the title, "people-sacred society." Whereas Becker's phrase conjures up the image of a long lost past, the new phrase intends to portend a revitalized future.[31]

The greatest danger that lies before us is less that the present secular order will survive many more decades, than that the process of resacrilization will be perverted. The danger is that we may move into what Becker called the "prescribed-sacred society". It is what the movement in the direction of the anti-communities seems to portend.[32]

The alternative, the people-sacred society, it if is to be achieved, must be built obliquely. Mankind will make the existing social order obsolete by rediscovering the sacred in the web of communal organizational bonds that link all human beings to each other.

---

[29] The term resacrilization is from Gerald Sykes who uses it precisely in the manner that is intended here, meaning, making sacred again. See, Gerald Sykes, "New York as Axis Mundi," *The Center Report,* (Santa Barbara, Center for the Study of Democratic Institutions) IV, 5, 1971, pp. 5-7. Also see, Abraham H. Maslow, *The Psychology of Science* (Chicago: Regnery, 1966) p. 138f.

[30] Howard Becker, *op. cit.*

[31] This search for a more stable, more ordered world is given effective expression by Irving Kristol, "When Virtue Loses All Her Loveliness", *The Public Interest* 21, Fall 1970, pp. 3-15. It is not entirely clear whether Kristol is expressing his own view, or whether he is observing and recording the view of others. For another but generally similar view see, Bertram M. Gross, "Planning in an Era of Social Revolution", *Public Administration Review,* 31,3, 1971, pp. 259-96.

[32] Becker took Nazi Germany as his prime example of the prescribed-sacred society. For a view of its possible application to the American scene see, Bertram M. Gross, "Friendly Fascism: A Model for America", *Social Policy,* 1,4, 1970, pp.44-53.

# GLOSSARY

Active generalized cooperation - mutual aid, friendship, neighborliness

Anti-community not a community - a type of social organization in which generalized cooperation on the family level is impossible

Coercive Power usually associated with physical force; restricting another's options to a desired *one.*

Commitment a. vertical - the sentiment of elements for the inclusive system b. horizontal - the sentiment of the elements for each other

Communal organization a relatively highly institutionalized social system not oriented toward a specific goal; a low goal oriented system

Community a local society, a communal organization including formal and communal subsystems

Community relevant elements the elements of the external system which are linked to a specific community

Cognitive elements the patterns and categories of thought supportive of a particular social system

Contract an agreement to limit cooperation

Crescive Community a type of historical community; a community which is low goal oriented on all three levels, e.g. a village, a city

Decommunization movement of a community in the direction of an anti-community

Democratization in a planned community, the gradual elimination of the staff-resident split

Elements individuals (in micro-systems) and/or groups (in macro-systems) which are the subsystems of an inclusive system

External linkage the relationship between an inclusive system and another system which is beyond the boundaries of the inclusive system

Formal organization a relatively highly institutionalized social system oriented toward a specific goal; a high goal oriented system.

Functional collectivity orientation the function of the elements with respect to the specific goal of the inclusive system

Generalized cooperation undefined and unlimited cooperation among the elements of a social system

Goal erosion in an Administered or an Intentional Community, when community level goals become less pronounced over time

Horizontal linkage the relationship between two or more included systems

Ideological goal similar to a specific goal except that it does not serve as an input into another *social* system

Macro-system a social system whose elements are subsystems

Mechanistic interaction for every action there is an expected, predictable reaction

Micro-system a social system whose elements are individuals, e.g., a family

Normative elements the patterns of behavior which are supportive of a particular social system

Normative emphasis the bias of a society in favor of particular social-structural patterns over others

Normative power the ability to control by symbolic means:horizontally - social controlvertically - persuasion

Official a leader who derives his legitimacy from his formal status within a system

Organization a relatively highly institutionalized social system

Partnership a contract between the two high goal oriented levels in a deviant community

Passive generalized cooperation laissez-faire, live and let live; may be horizontal or vertical

Power the ability to exercise control over the actions of others

Reciprocal partnership a partnership in which the major elements have reciprocal goals, as in a market system

Sentimental collectivity orientation loyalty, vertical commitment, i.e., the sentiment which the elements express toward the inclusive system

Social system a patterned social interaction which persists over time

Solidary partnership a partnership in which the major elements share a single, specific goal

Specific goal a measurable output of a social system that can serve as an input into another social system

Staff within Administered and Designed Communities, individuals who are the advocates of the goals of community relevant elements

Status position within a social system; in this sense, there are no degrees of status

Structural free wheeling change in one element does not produce necessary or predictable change in another element

Total cooperation a contractual agreement in which cooperation is unlimited, i.e., there is no opportunity not to cooperate

Utilitarian power the ability to control by means of the exchange of assets

Vertical linkages the relationships between inclusive systems and included systems, and vice versa

Vill a collective word for town, village, and city introduced by Hillery

# APPENDIX A

# THE THEORY OF COMMUNAL ORGANIZATIONS

The phrase communal organization is taken from Hillery's *magnum opus* of that title. Let us first, in brief outline, summarize the central thrust of Hillery's position. Later, as this analysis develops, we shall return to examine with greater specificity some of the details of this thesis.[1]

From an inductive analysis of fifteen case studies of a variety of communities, Hillery, following the folk-urban tradition in sociology, derives two modal types: the folk village and the city. Folk villages are defined in terms of three "foci": space, mutual aid, and families. Cities, by contrast, are defined in terms of the foci of space, contract, and families. Mutual aid and contract, suggests Hillery, are two forms of cooperation. Therefore, folk villages and cities constitute the two sub-classifications of what Hillery calls "vills." A vill is defined as "a localized system integrated by means of families and cooperation."

Hillery introduces the term "communal organization" to refer to a social system which is relatively highly institutionalized, but which lacks primary orientation toward a specific goal, in the Parsonian sense. Nations (not states), vills, neighborhoods, and families, are types of communal organizations. Although communal organizations may contain sub-systems that are specific goal oriented, or may have short-term, programmatic goals, they differ from other types of social systems, such as formal organizations, in that they are not directed toward a primary, defining, specific goal. Communal organizations are mutually inclusive,

---

[1] The following summary is based upon Hillery, *op. cit.,* as well as Hillery's own synopsis of his central concepts in George A. Hillery, Jr., *The Community Theories of Talcott Parsons,* paper read at the annual meeting of the Society for the Study of Social Problems, San Francisco, 1967, mineo.

i.e., they may be included within other communal organizations, or they may include them, vertically.

This, then, is the major thrust of Hillery's theory of communal organizations. Let us now examine it in greater detail.

### Communal Organizations as Social Systems

A communal organization is a type of social system. A social system is a basic unit of analysis in theoretical sociology. The elements of social systems are roles, or other social systems. Talcott Parsons uses the term social system with some lack of consistency. At times, social system refers to the total social system, which is coextensive with the entire society.[2] In these instances, the social system is theoretically to be contrasted with, e.g., the political system, or the economic system. However, Parsons makes relatively little systematic use of this meaning. Most commonly, he uses social system, to denote what he calls, "collectivities." He appears to use this latter word synonomously with the word organization. Organizations, according to Parsons, are social systems that have "primacy of orientation to the attainment of a specific goal."[3]

Collectivities (i.e., formal organizations) and societies are not the only types of social units to which the term social system is occasionally applied. Markets and market-like social structures, synthetic categories, i.e., analytical systems, and aggregates with at least one common identifying characteristic, are sometimes referred to as social systems.[4] Communal organizations resemble formal organizations in that they are both solidary and are interactional. By contrast, markets are not solidary, and analytical systems and aggregates are not interactional.

As Hillery suggests, the distinctive difference between communal organizations and formal organizations is that they are not oriented toward a specific goal, in the sense defined by Parsons. A specific goal is characterized by the following criteria:

1. The product of the goal is identifiable;
2. The product can be used by another system;

[2] "A total social system which ... may be treated as self-subsistent ... is here called a society." Talcott Parsons, *Toward a General Theory of Action* (Cambridge: Harvard University, 1951) p. 196.
[3] Talcott Parsons, *Structure and Process in Modern Societies* (Glencoe: Free Press, 1960), p. 17.
[4] Ramsoy, *op. cit.,* p. 173.

3. The product is amenable to contract, e.g., it can be bought or sold.[5]
Hillery adds to these criteria by suggesting that the products should be quantifiable or measurable.[6]

Communal organizations are similar to formal organizations in that they may vary in degree of inclusiveness and in degree of institutionalization (defined as, "the integration of complementary role expectations and sanction patterns").[7]

Hillery in his comparative analysis of formal and communal organizations does not take us beyond a discussion of the dimensions of goal-orientation, institutionalization, and inclusiveness. The two types of organization differ with respect to specific goal orientation. That is Hillery's main point. They are similar in that both are highly institutionalized. On this dimension, formal organizations differ from crowds in the same way that communal organizations differ from cliques.[8] Formal, like communal organizations, may vary not only in number of subsystems, but also in the number of *levels* of subsystems which they include.

Analytically, communal organizations appear to occupy a similar "social space" as Parsons' collectivies. They are limited in a similar manner by the "value aspect" and the "institutional aspect" of the total society.[9] As concepts, they exist on a similar, though relatively low level of generalization. Empirically, they not only interact with each other, but at times nearly coincide in their domains, as for example the town government and the town.

An exhaustive theoretical study would be required to identify and examine all of the characteristics of formal organizations, and to determine whether and to what degree they apply to communal organizations. We shall only begin this task here. In this study, we shall assume as a working hypothesis that communal organizations and formal organizations resemble each other in all respects except the five areas discussed below: specific goal orientation, linkage, the inclusion of subsystems, leadership, and collectivity orientation.

*Specific Goal Orientation*    Hillery insists that the difference between formal organizations and communal organizations is absolute: formal

[5] Parsons, *op. cit.,* 1960, p. 17.
[6] George A. Hillery, Jr. "Freedom and Social Organization: A Comparative Analysis," *American Sociological Review,* 36, February 1971, p. 54.
[7] Parsons, *op. cit.,* 1951, p. 191.
[8] Hillery, *op. cit.,* 1968, p. 147.
[9] Parsons, *op. cit.,* 1960, p. 170.

organizations have specific defining goals in the Parsonian sense, whereas communal organizations have no such goals. It would appear however, that the absolute distinction between the presence or absence of a
specific goal is difficult to reconcile with empirical reality. Take the
nuclear family as an example. The family, suggests Hillery, is a communal organization. But one of the critical functions of the family is
production and socialization of the next generation, i.e., a specific goal.
Some families are more highly specific goal oriented that others. For
example, patriotic families in Nazi Germany loyally produced sons for
the Führer. Other, probably most other families in the world, lack such
clearly defined goals. The point is, that families can differ in degree of
goal orientation.

Let us take another example. Hillery recently raised the question
whether religious convents are communal organizations. In his analysis
he divides convents into two types: apostolic and monastic. An apostolic
convent is involved in works in the larger society, e.g., education, while
a monastic convent is a contemplative religious community. Hillery
concludes that the apostolic convent is "an historic form of communal
organization which has been reorganized such that it operates as a
formal organization."[10] By contrast, monastic convents, according to
Hillery, are authentic communal organizations since they have no
specific goal. But, we respond, might it not be argued that salvation is
the specific goal of the monastic convent? Is it not a product, a specific
goal in the Parsonian sense, serving as an input into another system,
God? God knows and measures the results.

Two separate, and highly important issues have been raised in the
above two paragraphs: (1) the difference between the absence and the
presence of a specific goal appears to be a difference of degree, and (2)
there is some room for ambiguity and misunderstanding in the identification of specific goals by means of the criteria suggested by Hillery and
Parsons.[11] The second problem is largely resolved by changing the focus
of the analysis from the specific *goal,* i.e., the output, to the *orientation* of
the system. This idea has been discussed in greater detail in Chapter 9.

The first of the two problems may be resolved by suggesting that,
within the general classification of organizations there are two modes,

---

[10] George A. Hillery, Jr., "The Convent, Community, Prison or Task Force?", *Journal for
the Scientific Study of Religion,* VIII: I, 1969, p. 150.
[11] Hillery freely admits that the concept of specific goal is one with which his readers
frequently have difficulty. (Personal communication).

the formal and the communal. Each mode, each type of organization contains, in part, the characteristics of its opposite. The differences between the two types of organizations are not absolute, they are differences of degree along a continuum between high and low specific goal orientation. However, the distribution of organizations along this continuum is not in the shape of a normal curve; it is bimodal. Organizations that occupy middle positions, between the two modes, experience significant internal conflict, such as Hillery noted in the case of the monastic convent, between the "worshipful" and the "efficient".[12] We shall return to this point below.

*Linkage*  Hillery, in the early part of his volume, introduces the important concept of "structural free-wheeling," by which he means, "that change in one part of a communal organization does not mean that a mathematically predictable change must occur in another part of the system." He suggests that communal organizations are "linked by symbolic ties," that such links have different meanings for different people, thus accounting for structural free-wheeling.[13]

Hillery utilizes this important idea only in his analysis of the vill. The relationship between the three foci of the vill, space, family, and cooperation, is described in terms of structural free-wheeling: a change in one focus does not lead to a mathematically or mechanically predictable change in the other two.[14]

What are the foci, in a systematic sense? For Hillery, they are, "integrating constructs" which are "used in place of a definition," they are concepts that integrate the components of a theoretical model.[15] The foci are not interactional systems. Therefore, Hillery's use of the phrase structural free-wheeling within this context makes no direct reference to the linkages among interactional elements, i.e., the interactions of subsystems of vills in particular, or of communal organizations in general.

It is surprising that the analysis and application of the concept structural free-wheeling is not carried further by Hillery. Does he really intend to limit its use to a description of a relationship between analytical constructs, such as the foci of the vill? In introducing the

---

[12] *loc. cit.*

[13] Hillery, *op. cit.*, 1968, pp. 8,9.

[14] *Ibid.*, p. 69f.

[15] *Ibid.*, pp. 27, 30.

concept, he contrasts it with linkages which are more mechanistic, which emphasize mathematically predictable relationships. In these latter analyses, the mechanistic relationships generally refer to empirical, interactional systems, as well as to analytical systems. Therefore there is good reason to suggest that the concept of structural free-wheeling can be similarly applied in the case of communal organizations, to describe not only the nature of the relationship between the foci, but also the linkages among the interactional elements of the system.

Hillery, in contrasting the folk villages with the city, indicates that in folk villages the elements "cooperate by means of mutual aid," whereas in the city the elements are more likely to be "integrated by contract."[16] Both mutual aid and contract are forms of cooperation. Mutual aid is, "cooperation in general" (hereinafter referred to as, generalized cooperation) and contract is an agreement to limit cooperation.[17]

All forms of human cooperation, indeed nearly all forms of communication among humans, are dependent upon symbolic interactions. This applies equally to contracts and to generalized cooperation. The difference between the two types of cooperation is a difference of specificity. The symbolic language of contracts is able to limit the nature and the scope of interaction because it is a language which aims at maximal precision, based upon an institutionalized, socially agreed upon conceptual vocabulary. Whereas contracts deal with the assignment and specification of limited functions with respect to the goal of the organization, generalized cooperation permits of more latitude, and less precision. When philosopher Susanne Langer speaks of "man's need for symbolization," and "man's fundamentally expressive nature," her reference is, at least in part, to the type of symbolic interaction that is characteristic of this second, generalized type of cooperation.[18]

Contracts usually specify exchanges between or among cooperating units. Each party both gives and receives something. By contrast, in generalized cooperation, where exchanges also take place, they are not specified and are likely to be random. There is no specific exchange of this for that; there is structural free-wheeling without mechanistic interaction.

---

[16] *Ibid.,* p. 51f.

[17] *Ibid.,* pp. 38, 54; also see Pitrim Sorokin, *The Crisis of Our Age* (New York: E. P. Dutton, 1941) p. 167f.

[18] Susanne Langer, *Philosophy in a New Key* (New York: New American Library, 1942) pp. 33-41. Also see, Scott Greer, *The Logic of Social Inquiry* (Chicago: Aldine, 1969) p. 37f.

The exchanges which take place as a result of contractual agreements usually involve measurable resources such as money, specified powers, or material objects.[19] Something that cannot be measured is, after all, difficult to specify within a contract. Within generalized cooperation such measurable resources may also be exchanged. But more important to generalized cooperation, are exchanges of non-measurable resources, e.g., love, knowledge, tradition, or sentiments. "Resources" such as these are not limited in the same way that measurable resources are limited. There is no limit to love, to knowledge, or to culture; the more that is produced or acquired by any element of the system, the more exists. In short, the internal economics of communal organizations within which generalized cooperation prevails is significantly different from the economics of formal organizations.

Let us make explicit the conceptual relationship between structural free-wheeling and generalized cooperation. Both are here used to refer to the linkages among the interactional elements of communal organizations. Both have their roots in non-specific, symbolic interactions. Generalized cooperation is to contract, what structural free-wheeling is to mechanistic interaction. The first set of terms gives a name to and defines the scope of the interaction; the second set describes the nature of the interaction (See Plate 12).

## PLATE 12

### THE PRIMARY TYPES OF COOPERATION AMONG THE ELEMENTS OF ORGANIZATIONS

|  | Formal Organizations | Communal Organizations |
|---|---|---|
| The Scope of Cooperation | Contract (agreement to limit cooperation) | Generalized cooperation; 1. Active generalized cooperation (mutual aid) 2. Passive generalized cooperation (laissez-faire) |
| The Nature of Cooperation | Mechanistic interaction | Structural freewheeling |

---

[19] See, Peter Blau, *op. cit.*

Hillery equates generalized cooperation with mutual aid. Mutual aid, in every-day language is used to refer to an active, assertive, explicit helping relationship between units. We would observe, however, that this is only one form of generalized cooperation.

Generalized cooperation also has a passive, less assertive, implicit mode. It refers to a relationship of live and let live, of laissez-faire, as well as one of mutual aid. Thus, an implicit agreement, such as may exist between two neighbors to minimize communication and to respect each other's privacy, is also a form of generalized cooperation, though it can hardly be called mutual aid.

In order to gain a further insight into the nature of the linkages among the elements of communal organizations, let us return to the discussion of the vill. Among the interactional subsystems of the vill, the terms linked, cooperate, and integrated, all refer to the same thing, the *horizontal linkages* among the subsystems. The systematic difference between the city and the folk village lies in the differential nature and scope of their horizontal linkages. In cities, formal organizations are more prevalent. Since formal organizations are linked by means of contract, the difference between the city and the folk village is that in the city horizontal linkages are more frequently by contract, whereas in the folk village, they are usually by generalized cooperation.

Horizontal linkages between two communal organizations within a vill, e.g., two families, are most commonly active or passive generalized cooperation. Horizontal linkages between two formal subsystems of the vill, e.g., two local business enterprises, are most commonly by contract. The horizontal relationships between included communal organizations and included formal organizations are of special interest, e.g., the linkage between families and businesses, or between families and governmental organizations. Such linkages may be either by contract or by generalized cooperation, or both.

Eugene Litwak's analysis is of relevance here. Litwak has identified eight "mechanisms" which serve to "coordinate" the two types of organizations: detached experts, opinion leaders, settlement houses, voluntary associations, common messengers, mass media, formal authorities, and delegate organizations.[20]

---

[20] Eugene Litwak, "A Balance Theory of Coordination Between Bureaucratic Organizations and Primary Groups", *Administrative Science Quarterly*, 11:31-58, 1966; also see, Eugene Litwak, "Technological Innovation and Functions of Primary Groups and Bureaucratic Structures", *American Journal of Sociology*, 73: 471, 1968.

Litwak's mechanisms are a mixed bag. His theory requires more detailed discussion than can be given it in the resent context. Litwak speaks not of communal organizations but of primary groups. Whereas there is much overlap between the two analytical types, it is essential to recognize that all primary groups are not communal organizations and that all communal organizations are not primary groups.

But, with this difference aside, the important additional insight to be gained from Litwak's analysis is that horizontal linkages between formal and communal organizations in vills cannot be fully explained by means of the contracts and the generalized cooperation between them. Frequently such linkages require the interposition of specialized mechanisms to facilitate the interaction.[21]

*Horizontal linkages* are to be contrasted with *vertical linkages.*[22] By vertical linkages are meant the relationships between the subsystem and the inclusive system. (It is important that the term, vertical linkage, be distinguished from Roland Warren's "vertical patterns," by which he means the relationships of community subsystems to extra-community systems. We call these, *external linkages.*)[23] In formal organizations vertical linkages are primarily contractual, and of great significance. In communal organizations the relationship between included systems and the inclusive system may best be characterized by what has been referred to above as passive generalized cooperation.

Both the folk village and the city, i.e., all vills, have in common the negative fact that they are not vertically linked by contract. This is simply another way of pointing to the fact that vills, as communal organizations, are low goal oriented. Their capacity to act in pursuit of a specific goal is low because their subsystems, families, ethnic groups, corporations, etc. are not linked to it (i.e., to the vill) by contract.

In this connection it is important to distinguish, as Hillery does, between the vill and its government. The government is a formal organization which both embodies and regulates contractual links within the vill. The government of the vill is not to be equated with the totality

---

[21] Litwak proposes that the mechanisms which he has identified provide coordination—by which he means influence and/or control—in either direction. But he fails to prove his point. His discussion is limited to an analysis of coordination from bureaucratic organization to primary group. The reverse, coordination from primary group to bureaucratic organization, is, we suspect, achieved largely by different means.
[22] It is tempting to use linked for vertical processes, and integrated for horizontal processes. But that would only add to present complications.
[23] Roland Warren, *The Community in America* (Chicago: Rand McNally, 1963) p. 161.

of the vill. It is one of its formal organizational subsystems, and, it might be added in passing, not always the most important.

Warren's emphasis upon vertical patterns, i.e., external linkages, is not central to this discussion because it refers to linkages between formal organizations, not communal organizations. Formal organizations, as indicated, have strong vertical contractual links. In North American society, the formal organizational subsystems in communities, e.g., corporations and political parties, are usually externally linked to inclusive level formal organizations which are located beyond the systemic boundaries of the local society. However, the vertical links of the included formal subsystems in the community *to the community* are of the same nature as the vertical links of the communal subsystems— they are linked by means of passive generalized cooperation.

Plate 13 diagramatically summarizes the differences between formal organizations and communal organizations with respect to their linkages. Formal organizations are linked primarily be means of contract, whereas communal organizations are linked primarily by means of generalized cooperation. Whereas the nature of the linkage of formal organizations tends to be mechanistic, the linkage of communal organizations is characterized by structural free-wheeling. When the inclusive system is a communal organization, as in the case of the vill, vertical linkages are characterized by passive generalized cooperation. This does not, however, preclude included formal organizations from being linked horizontally, and/or externally, by means of contract. Included communal organizations are linked horizontally to each other by either passive or active generalized cooperation. They are linked horizontally to formal organizations either by contract or by generalized cooperation.

*The Inclusion of Subsystems*    The elements of communal organizations, like all interactional social systems, are roles and subsystems. Communal organizations include formal as well as communal organizational subsystems, e.g., communities include corporations as well as families. The question arises, whether the same relationship holds between inclusive level formal organizations and their subsystems, formal organizations similarly include not only formal, but also communal organizational subsystems. Closely connected with this issue is a second question. If communal organizations are included within formal organizations, what is the nature of the control exercised by the inclusive system over the included system?

## PLATE 13

### *THE LINKAGES OF FORMAL AND COMMUNAL ORGANIZATIONS, CITIES AND FOLK VILLAGES*

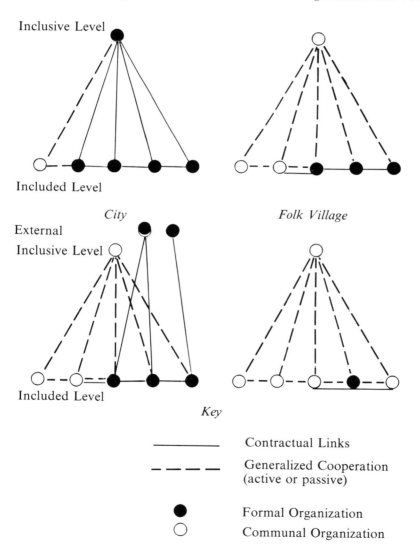

*Formal Organization*

*Communal Organization and Vill*

Inclusive Level

Included Level

*City*

*Folk Village*

External
Inclusive Level

Included Level

*Key*

| | |
|---|---|
| ———————— | Contractual Links |
| — — — — | Generalized Cooperation (active or passive) |
| ● | Formal Organization |
| ○ | Communal Organization |

According to the "classical theory" of administration, the efficiency of a formal organization is maximized when roles and functions are clearly defined with respect to the organizational goal and there is a minimum of interference in this process from so-called informal systems.[24] According to this view inclusive level formal organizations include formal subsystems, but tend to strongly discourage the formation or the inclusion of communal subsystems. For example, bridge games at the office are discouraged, or prohibited.

This view has been challenged by the "human relations approach," which gained its first significant impetus from the famous Hawthorne studies of the early 1930's.[25] These, and other similar studies which followed, tended to indicate that included informal systems contribute to the goal achievement of formal organizations. These findings led to the general conclusion that the humanly more satisfying is also the more efficient. It is essential to this view, however, that the inclusion of a communal subsystem within a formal organization is functional, and therefore desirable, only as long as it is under the ultimate control of, and/or subservient to the goals of, the inclusive formal organization.

The more recent, structuralist approach to the analysis of human organizations views the formal organization from a multiplicity of perspectives and emphasizes other functions of the organization, in addition to specific goal attainment.[26] Where and if the structuralist analysis leads to practical implementation, the result is an inclusive level formal organization that begins to resemble a communal organization. Within such a setting, communal organizational subsystems may exist side by side with formal organizational subsystems, not necessarily subservient to the specific goal of the inclusive organization. The American university might be an example of such a system. On the industrial level, the Scott Bader Commonwealth in England is an example.[27]

In sum, it is evident that both formal and communal organizations may include both types of subsystems. However, the more highly goal oriented the inclusive system, the less likely it is to sanction communal subsystems which are not subservient to it. Subservience within this

---

[24] Amitai Etzioni, *Modern Organizations* (Englewood Cliffs, N.J.: Prentice Hall, 1964) p. 22f.

[25] F. J. Roethlisberger and W. J. Dickson, *Management and the Worker* (Cambridge: Harvard University Press, 1939).

[26] Etzioni, *op. cit.,* 1964, p. 41f.

[27] Fred H. Blum, *Work and Community* (London: Rutledge and Kegan, Paul; 1968).

context means, quite specifically, subject to the control of and the limitations set by the inclusive system.

*Leadership*   The hierarchical structure of formal organizations is too familiar to require extended discussion. Within such organizations it is the function of formal leaders, i.e., officials, executives, etc., those who are near the top of the hierarchy, to exercise the greatest organizational control. Others, individuals and subsystems further down the hierarchy, share in the exercise of organizational control, but to a lesser degree.

Communal organizations are non-hierarchical. Their leaders tend to be informal, personal, and charismatic, and their primary source of legitimacy is not derived from their status within the organization. However, communal organizations do allow for differences in status and role among their elements. Status, in this connection, refers only to position, and not to privilege or to superior power within a hierarchy.[28] A person within a family, for example, may have the status and fulfill the role of father, or of son. This does not locate him within a stratified hierarchy of power or privilege, except in the case of an authoritarian family. Specific perogatives and obligations are attached to each status within a family, but this is not to be confused with privileged entitlements and superiority, as in a hierarchy. Ideally, all members of a family, though occupying different statuses, are in this sense equal.

Similarly in the vill, there is a variety of statuses and roles among subsystems, but no single formal hierarchy. Norton Long speaks of an "ecology of games" within a community, i.e., an array of subsystems, some formal and some communal, which are but loosely coordinated.[29] Floyd Hunter's study of the power structure of Atlanta reveals the existence of "crowds" among the economic elite.[30] These crowds are organized along lines of friendship and common interest, and do not necessarily conform to the hierarchical lines defined by the world of business. The crowds represent the tops of hierarchies, but they are informally and non-hierarchically linked to each other.

---

[28] We follow Linton's early usage, not Weber as translated by Bendix. Status means position in a social system. *Vide,* Ralph Linton, *The Study of Man* (New York: Appleton-Century, 1936) pp. 113-114. Also see David Gil, "A Systematic Approach to Social Policy Analysis," *The Social Service Review,* 44, 4, December 1970, p. 417.
[29] Norton Long, "The Local Community as an Ecology of Games," *American Journal of Sociology,* LXIV, November 1958, pp. 251-261.
[30] Floyd Hunter, *Community Power Structure* (Chapel Hill: Univ. of North Carolina Press, 1953) p. 77f.

It is possible that a communal organization has no identifiable informal leaders, or that it has multiple leaders, none of whom are in competition with each other for organizational control. Who, for example, leads in a genuine friendship? Many modern communes, organizations which are best understood if one thinks of extended families, conscientiously deny the existence of leadership. They may have task leaders, but these are individuals performing specialized roles and are not to be considered organizational leaders. In ethnic groups informal or charismatic leaders seem to come and go. Only by public opinion poll might it be possible to identify this year's five leading American Jews, or Irishmen, or Blacks.

Amitai Etzioni has suggested that organizational control may be divided into three types: physical, material, and symbolic. Each of these types is applied by a parallel type of power: coercive power, utilitarian power, and normative power.[31] Power is defined most simply as, the ability to exercise control over the actions of others. It is exercised through the utilization of positive and negative physical, material and/ or symbolic sanctions. Coercive power refers to the threat or exercise of force, usually physical force, to impose a desired course of action upon others. Utilitarian power, by contrast, refers to the exchange of assets between units for the purpose of gaining support for a desired course of action. Utilitarian power may be exercised either by the granting or the withholding of assets. Normative power, when exercised horizontally, e.g., among peers, may be equated with social control. When normative power is exercised vertically, by an inclusive system over an included system, or vice versa, then it becomes persuasive power, as for example the rallying of public opinion by means of propaganda.

All social systems are controlled primarily by normative power, based upon a general consensus among all elements concerning good and desirable behavior and action. In formal organizations such power, operating both vertically and horizontally, is heavily relied upon by leaders in their exercise of control, *but in conjunction with utilitarian and coercive power.* One of the important differences between formal and communal organizations lies in that, in formal organizations all three types of power may legitimately be exercised, whereas in communal organizations only normative power is legitimate.

---

[31] This discussion is based upon, Amitai Etzioni, *A Comparative Analysis of Complex Organizations* (New York: Free Press, 1961) and Amitai Etzioni, *The Active Society* (New York: Free Press, 1968) pp. 355-357.

In formal organizations, leaders depend primarily upon normative power, secondarily upon utilitarian power, and only as a last resort upon coercive power.[32] It is one of the attributes of leadership in communal organizations that it is entirely dependent upon normative means to influence the system of which it is a part. There is evidence that in those instances where informal leaders successfully threaten the authority of formal leaders, the formal authorities are linkely to expand their use of coercive power.[33]

As noted earlier, communal organizations are linked by generalized cooperation, a type of linkage based upon non-specific, symbolic inter-action. Normative power is similarly exercised, by symbolic means which are often non-specific, e.g. reminders to do good, or to follow tradition. The absence of formal hierarchy in communal organizations is connect-ed with the fact that informal leaders, as well as all others, are limited to the utilization of normative power. An informal or charismatic leader is precisely that kind of person who has the ability to persuade without explicit or implicit resort to coercion or bribery. In the event that an informal leader does exercise utilitaran or coercive power, he is in danger of undermining his own status as a leader. It is like the parent who beats his children and in the process loses their respect.

In communities, informal leaders are sometimes invited to join formal leaders in decision making bodies. Poverty programs, and many other governmental and voluntary agencies in the social welfare field have utilized this method of gaining representation from "the poor." However, when informal leaders begin genuinely to participate in these formal decision making structures then their legitimacy as informal leaders is in danger of being called into question.[34] Either they may be cut off from their constituency (because they no longer exercise exclusively normative power), or they are in the process of converting their constituency into a formal organization, e.g. politicizing it and thereby imposing upon it specific goals.

---

[32] Even the most coercive societies, prisons, depend primarily upon non-coercive power for social control. See, Sykes, *op. cit.* pp. 40-62.

[33] Etzioni calls this a "trade off curve" between control and consensus: the more consensus, the less the need for coercive control from the leaders. *Vide,* Amitai Etzioni, "Toward a Theory of Societal Guidance," *American Journal of Sociology,* 1967, p. 182. The same point is made more dramatically by Hannah Arendt, who argues that, as states and empires begin to feel their control slipping, they increasingly resort to violence, i.e., coercion. Hannah Arendt, *On Violence* (New York: Harcourt Brace, 1969).

[34] On this theme, see, Peter Marris and Martin Rein, *Dilemmas of Social Reform* (New York: Atherton, 1969), esp. pp. 164-190.

This last point leads to a related issue to which we alluded earlier. What happens when inclusive level formal organizations become excessively coercive from the point of view of an included communal organization? *Must* included communal organizations be subservient, or do they have another choice? It would seem that they have a total of three choices: to remain subservient, to "organize," or to resist normatively. To organize means that they begin along the road of conversion, to becoming a formal organization. That is what happens when a neighborhood or an ethnic group becomes politicized. The choice of normative resistance does not require such a conversion. Normative resistance involves a challenge to the legitimacy of the inclusive system utilizing methods such as noncooperation and nonviolent resistance. This is what happens, for example, when families refuse to send their sons to war, or when the Czechoslovakian people deride and shame the Soviet soldiers in their tanks as they invade the city of Prague.[35]

In summarizing this section on leadership, we conclude that formal organizations are hierarchically organized, and that they have formal leaders who exercise coercive, utilitarian, and/or normative power. By contrast, communal organizations lack formal hierarchies and are controlled by normative power which is exercised by peers (i.e., by social control) and/or by informal leaders (i.e., by persuasion). In a contest between formal and communal organizations, formal organizations use all three kinds of power; communal organizations, if they are to remain communal organizations, will rely on normative power.

*Collectivity Orientation*    Parsons defines collectivity orientation as, "a state of positive membership whereby the norms and values of the higher order system (i.e. the inclusive system) are partially prescriptive for the action of the low (the included system)."[36] Parsons suggests that collectivity orientation is a variable which describes an aspect of every institutionalized role. When failure to identify with the inclusive system threatens the integrity of that system then a "moral issue" is raised, the issue of loyalty. "Collectivity orientation, as it were, involves posing the 'question of confidence,' are you one of us? Your attitude toward the question decides."[37]

---

[35] Hannah Arendt, *op. cit.,* pp. 52-53.
[36] Talcott Parsons and Neil J. Smelser, *Economy and Family* (London: Routledge and Regan Ltd., 1956) p. 36.parenthesis added.
[37] Talcott Parsons, *The Social System* (New York: Free Press, 1951) p. 97.

To the extent that collectivity orientation is associated with expressive loyalty it is called sentimental collectivity orientation. Sentimental collectivity orientation is a variable which applies to both formal and communal organizations. It is misleading and probably empirically inaccurate to assert that the elements of communal organizations are more likely to exhibit sentimental collectivity orientation (sometimes called commitment) than the elements of formal organization. This negative point is important to the extent that it is in disagreement with the position taken by proponents of the "community movement," those who lament the loss of the "sense of community," the loss of loyalty to the locality which is characteristic of modern industrial society.[38]

Sentimental collectivity orientation may be conceptually separated and distinguished from functional collectivity orientation. Whereas the first refers to expressive loyalty, the second is simply that aspect of the role of an included element which relates to the inclusive system. In communal organizations functional collectivity orientation is minimal; the elements are functionally highly autonomous of the inclusive system. In formal organizations, by contrast, the principal functions of the elements consist of those actions which refer directly to the inclusive system and its specific goal. In such organizations, functional collectivity orientation is the primary source of integration of the system. This distinction may be expressed in another way: whereas formal organizations are to be understood primarily in functional terms, communal organizations are not primarily functional, but existential.

This leads to another distinction between formal and communal organizations. Whereas in formal organizations the roles of the elements are defined primarily in terms of their relationship to the inclusive system, in communal organizations the reverse is true: the system is defined by the elements. This requires explanation and illustration.

Formal organizations are usually elaborated from the top down. Bureaucratic expansion consists most commonly of the definition and institutionalization of new functions within subsystems. Whereas, in formal organizations, the addition or the redefinition of subsystems is readily justified in terms of the specific goal, the redefinition of roles and functions at the top by those on the bottom is less common. In the extreme case, it is tantamount to revolution.

---

[38] Vide, e.g. Robert A. Nisbet, *Community and Power* (New York: Oxford, 1953); Maurice R. Stein, *The Eclipse of Community* (New York: Harper, 1960); Lawrence Haworth, *The Good City* (Bloomington: Indiana University, 1963). Also see Chapter 10, *supra*.

It might be suggested that associations, as a type of formal organization, are an important exception to the rule that role definition takes place from the top down. Associations are inclusive level formal organizations created by previously independent actors in the pursuit of a common goal. In such organizations, the role of the inclusive level is initially defined by the included subsystem—but not for long, as Mancur Olson has shown. In high goal oriented systems such as associations, the inclusive level must quickly make itself largely independent of its constituent elements in order to be able to function effectively in pursuit of the collective goal.[39]

As mentioned in passing above, communal organizations are established within, and limited by, the same "value and institutional aspects" of society as formal organizations. Thus, for example, in America a family-like communal organization consisting of two homosexual adults is generally disdained, if not legally proscribed. The cultural form for family or for community preexists each specific instance of a family or a community. Beyond these broad cultural prescriptions, however, each communal organization derives directly from the action of its subunits. It is defined by its elements. The actors (individual and systematic) precede the roles, rather than the reverse, as is usually the case in formal organizations. Formal organizations are commonly created by external systems, and sometimes, as in the case of associations, they are generated by their elements. By contrast, communal organizations *must* be generated by their elements.

The family is an important case in point. Two previously independent actors are joined in marital union to create the family. Ideally, at least in Western society, families generate themselves; they are not a product of an inclusive level family-making organization. There are, of course, traditions and norms which influence the initial selection of the marriage partners. That is where the "value and institutional aspects" of society come into play. But each specific family is generated by its elements, by a man and a woman.

In a very real, physical sense, the children are created by the marriage partners. Obviously, they did not originally participate in the creation of the definition or the family. However, ideally and largely in practice, the family is regenerated, as it were, with the addition of children. In our society, as the children grow, and for as long as they remain members of

---

[39] Mancur Olson, *The Logic of Collective Action* (New York: Schocken, 1968).

the nuclear family, they continually attempt to redefine it qualitatively as something different from the original marriage partnership. Needless to say, more hierarchical, more coercive, more highly goal oriented families can be expected to follow a different pattern.

Nations and communities, like families, are communal organizations and as such are similarly defined by the actions of their subsystems. Families and other formal and informal organizations on the family level generate communities, which in turn may become the building blocks of nations. The roles of the subsystems with respect to the inclusive system are defined primarily by the incumbents, not by the inclusive system.

Formal organizations are integrated functionally, as a direct concomitant of their orientation toward a specific goal. As indicated, such functional integration may be accompanied to a greater or lesser degree by sentimental collectivity orientation. The question now arises, what is the basis of the integration of communal organizations? Among such organizations functional collectivity orientation is minimal, for the obvious reason that they have no specific goal. But what integrates a communal organization in the event that sentimental collectivity orientation also is minimal?

Symbols of identity serve to integrate communal organizations in the absence of sentimental collectivity orientation. The most common symbol of identity is a proper name, e.g., Americans, Negroes, the Kellys, etc. Physical attributes or objects may also serve as symbols of identity, e.g., skin color, a territory, or a language. It is not that such symbols of identity are unique to communal organizations, but rather that within such organizations, in the event of low sentimental collectivity orientation, they perform an especially important function.

The symbols of identity do not need to be conscious or explicit from the point of view of the elements of the system. Take, for example, a person who is a member of a community even without expressing loyalty to it, and without giving active consent to his membership. He has defined his own role with respect to the communal organization as one of passive generalized cooperation. Such an individual may be called a latent member of the inclusive system. In this case, the community with a name serves as little more than a point of reference, an identity. If a communal organization includes a large number of such latent members it does not necessarily cease to be a communal organization, nor does it become a communal organization with few members. Rather, it is simply a communal organization with low sentimental collectivity orientation.

Let us summarize. There is no difference between formal and communal organizations when only sentimental collectivity orientation is considered. However, with respect to functional collectivity orientation there is a difference: formal organizations possess high functional collectivity orientation and communal organizations do not. Another important difference between the two types of organizations is that in formal organizations the inclusive system defines the roles of its elements, whereas in the communal organizations the system is defined by the elements.

### An Alternative Conceptual Orientation

With all that has been said, contrasting formal with communal organizations, the picture is still not complete. We return to the awareness that some, though not all, communal organizations appear to occupy the same "social space" as formal organizations, as for example, nations and states, communities and municipal corporations. It may be suggested that in these cases, making the distinction between communal and formal organizations is more like talking about two different *aspects* of the same human group, than about two separate social systems. According to this view, communal and formal organizations which are parallel, such as nations and states, constitute simply a more emphatic institutionalization of what Homans calls the "internal system" and the "external system."[40] Homans' use of the word "system" in this connection if unfortunate because it is confusing. The external and internal are, from Homans' viewpoint, more accurately two aspects of a single system:

> We shall not go far wrong, if, for the moment, we think of the external system as group behavior that enables the group to survive in its environment and think of the internal system as group behavior that is an expression of the sentiments towards one another developed by the members of the group in the course of their life together.[41]

Let us carefully examine this parallelism between Homans' external and internal system, and our formal and communal organization.

Homans' analysis of small groups revolves around three concepts: activity, sentiment, and interaction. Activity is what individual members of the group do, sentiment has reference to the feelings possessed by

---

[40] George C. Homans, *The Human Group* (New York: Harcourt Brace, 1950).
[41] *Ibid.*, pp. 109-110.

individual members of the group, and interaction refers to their relationship with each other. Homans does not present his findings precisely in the manner in which we shall rearrange them. We will take each of his analytical concepts and identify the external and the internal dimension. To the extent possible we will rely on Homans' own language and case examples.

The *activity* of a group is, on the one hand, those actions which are dictated by the demands of its environment. In Homans' case example of the Bank Wiring Room, these demands derive primarily from the necessities imposed upon the workers by the General Electric Company.[42] This, then, is the external dimension of the activity of the group. The internal dimension of activity within the Bank Wiring Room consists of the friendly, mutual helping efforts which spontaneously arise among the workers. In this case the internal dimension of activity grows out of the external. By contrast, in Homans' case example of Norton Street Gang—a street corner club and therefore a communal organization—the internal dimension of activity predominates and the external dimension is hardly discernable. The Norton Street Gang is, what Homans calls an "autonomous" group.[43]

There are two kinds of *sentiment:* the sentiment which derives from the individual goals of the factory workers and the production goals of the group, and secondly, the sentiment which develops among the workers for each other.[44] Again in this case, if we contrast the Bank Wiring Room with the Norton Street Gang, in the former, goal oriented sentiment takes precedence. In the latter, the sentiment which is given expression among the members of the group constitutes the very essence of that group.

Though Homans does not use the terms expressive and instrumental, they most clearly distinguish between the two types of *interaction* which he infers. In the Bank Wiring Room, instrumental interaction and activity that is goal-directed are almost inseparable. Expressive interaction here "elaborates itself" out of the instrumental.[45] In the case of the Norton Street Gang, expressive interactions predominate, and instrumental interactions are almost unidentifiable. Here expressive interac-

---

[42] *Ibid.,* p. 96.
[43] *Ibid.,* p. 175.
[44] *Ibid.,* p. 118.
[45] *Ibid.,* p. 109.

tion and the sentiment of one group member for another reinforce each other.[46]

What we have accomplished in the above several paragraphs is to use Homans' analysis of two groups to elicit many of our distinctions between formal and communal organizations. The Bank Wiring Room is a formal organization where high goal oriented activities predominate and where linkages tend to be contractual. The Norton Street Gang is a communal organization which is low goal oriented and where generalized cooperation prevails.

Homans' analysis provides support for our contention that in formal organizations the roles of the elements (the activites, sentiments, and interactions of the members of the group) are defined by the inclusive system, whereas in communal organizations the inclusive system is defined by the elements. For Homans, in the case of the Bank Wiring Room, "the internal system arises out of the external."[47] In the case of the Norton Street Gang, "the increasing interaction itself is a starting point for the formation of the group."[48] The members define their own roles with respect to the group, and the external system is of little significance.

Homans' purpose is to suggest the analytical similarity of many types of small groups, rather than the systematic differences between two types of groups, as in our analysis. In discussing leadership of groups he suggests that all groups are hierarchical, without giving much attention to the difference between the hierarchies which serve an instrumental purpose, and expressive leadership. But the role of Doc, the leader of the Norton Street Gang, is not to be paralleled neatly with that of the supervisor of the Bank Wiring Room. Doc's position is entirely dependent upon his personal charisma and the group's assent. By contrast, in the Bank Wiring Room, leadership is formally hierarchical, imposed from above. The personal response of the workers to their supervisor is of a relatively low order of importance. The supervisor exercises power primarily by utilitarian means; Doc exercises almost exclusively normative power. Although Homans provides the evidence for these distinctions, he only partially develops them. He chose to emphasize the similarity of the presence of leadership statuses within the two groups, rather than the essential difference in the nature of this leadership.

This extended discussion of Homans' analysis of two groups serves not only to support the theory of communal organizations, but also suggests

---

[46] *Ibid.,* p. 112.
[47] *Ibid.,* p. 151.
[48] *Ibid.,* p. 173.

how, in the analysis of empirical phenomena, the theory may be utilized with a "split-level" approach. *Within* organizations the method of analysis is most similar to that used by Homans, identifying the high and low goal oriented aspects of the system. However, *among* organizations the distinction between formal and communal organizations is most appropriate.

### Formal versus Communal Organizations

The external-internal perspective suggested by Homans leads to an insight which might not have been as easily derived otherwise. Parsons, Bales, and others who would translate external-internal into adaptive-integrative, suggest the existence of a "series of reciprocal strains" between these two aspects of a system.[49] Similarly, the likelihood of strain between the elements leading to high goal orientation and the elements leading to low goal orientation within an organization might be suggested. This would help explain the potential for organizational change from formal organization to communal organization, and vice versa. These strains, this contest, are paralleled on the inclusive level, between formal and communal organizations. The contest on both levels is a way of accounting for the bimodal distribution of organizations along the continuum of high-low goal orientation. Organizations that occupy the middle position are likely to be more highly conflicted and experience greater strain.

The current dilemma of many American institutions of higher learning may be taken as an example of organizations attempting to straddle the middle position. For many, the university serves primarily as a means to an end; others prefer to discover within the university an alternative way of life. American Blacks are today in a similar middle position. Some leaders would organize Blacks along highly goal oriented, highly political lines. Others, such as the Muslims, reject politics and envision a black cultural renaissance.[50] Both the universities and American Blacks, can be expected to become increasingly polarized, according to this hypothesis.

---

[49] Robert F. Bales, "Adaptive and Integrative Changes as Sources of Strain in Social Systems", in *Small Groups, A. Paul Hare, Edgar F. Borgatta, Robert F. Bales, eds. (New York: Knopf, 1955) pp. 127-131.*

[50] For the moment, the Muslims appear to constitute a social movement with a cultural-religious mission which ideologically rejects participation in secular, i.e., "white" politics. It is probable that when and if a separate Muslim state is established, it will shortly begin to resemble other states. There are numerous examples in Western history of how social movements thus become formally institutionalized. On this topic see, Neil J. Smalser, *The Theory of Collective Behavior* (New York: The Free Press of Glencoe, 1963).

We conclude, therefore, by suggesting that there are contests both between formal and communal organizations, and within organizations, between their formal and their communal aspects. As with every contest or conflict, within it reside important elements of cooperation between the two opposing poles. Without such cooperation the survival of society is inconceivable. On the other hand, cooperation without contest serves primarily to perpetuate the *status quo.*

### A Summary of Major Definitions and Propositions

1. Human organizations may be divided into two classes: formal organizations and communal organizations. Formal organizations are characterized by high goal orientation and communal organizations by low goal orientation.

2. The difference between formal and communal organizations is one of degree, but the distribution of organizations along the continuum of goal orientation is bimodal.

3. Formal organizations are primarily linked by means of contract (i.e., agreements to limit cooperation), whereas communal organizations are primarily linked by means of generalized cooperation.

4. The linkage of formal organizations can best be described in mechanistic terms; the linkage of communal organizations can best be described in terms of structural free-wheeling.

5. Both formal and communal organizations include formal and communal subsystems.

6. Formal and communal organizations which occupy a similar "life space," or domain, are frequently in contest with each other. Similarly, within organizations, the formal aspects are often in contest with the communal aspects.

7. Inclusive level communal organizations which are vertically linked by means of passive generalized cooperation, may include formal subunits which are horizontally linked to each other and to included level communal organizations by means of contract.

8. Inclusive level formal organizations attempt to maximize their control over their formal as well as their communal subsystems, in the service of their specific goal.

9. Formal organizations are hierarchically organized and controlled by formal leaders who exercise coercive, utilitarian and/or normative power. Communal organizations lack such hierarchies and are controlled by means of normative power exercised by peers and informal leaders.

10. If the control of inclusive level formal organizations is contested by included communal organizations attempting to exercise coercive or utilitarian power, they risk being converted into formal organizations.

11. If inclusive level formal organizational leaders feel threatened by their included communal organizations they are likely to increase their exercise of coercive power.

12. Inclusive level formal organizations define the roles of their subsystems; inclusive level communal organizations are defined by their subsystems.

13. Formal organizations either are created by other formal organizations or they are generated by their elements. Communal organizations must be generated by their elements.

# COMMUNITY SCHEDULE

I.  History
    A.  Who took the first step toward the creation of the community?
    B.  What has been the subsequent role of these initiators and their heirs?
    C.  What are the most significant milestones in the development of the community?
    D.  What changes have taken place in the formal structure of the community over the years?
        1.  statement of purpose or purposes
        2.  legal status and political structure
        3.  boundaries and location
        4.  economic base
        5.  membership: nature and size

> *Note:* Each of the following items is to be examined historically i.e., what are the facts today, what changes have taken place over time, what were the facts at the time of the creation of the community?

II. Community Level Specific Goals
    A.  Is there an explicit or implicit statement of goals? What is it and what is its origin?
    B.  Does the goal of the community include a product which may serve as an input into another system? Is the product measurable? Is it amenable to contract?
    C.  Are there specific, identifiable steps which lead to the attainment of the goals of the community? Is there a specific process or technology? Give examples.
    D.  Are there specific, identifiable negative steps which lead away from the goals? Give examples.
    E.  How do individual members or families contribute to the attainment of the goals of the community?
    F.  Are organizations external to the community essential to the community in the attainment of its goals? Which ones, and how?

III. Initiation of Members and Role Allocation
   A. Is there a formal process of initiating new members into the community? Who administers and who controls this process?
   B. How are roles allocated to new members? By whom?
   C. Is there a method whereby roles are rotated among members? Who controls this process? Who would be authorized to make changes in the system?
   D. By whom are the membership boundaries of the community defined, i.e., who is considered a member, what kind of person may become a member, how large the total number of members should be?
   E. Is there disagreement among groups of members, or between members on the one hand and leaders on the other, concerning the principles of membership or the allocation or roles? What is the nature of this disagreement?

IV. Leadership
   A. Are there identified or identifiable community leaders? How do the leaders demonstrate their leadership?
   B. How are leaders chosen from among the members (residents) of the community? Is there a formal selection process or are leaders identified informally?
   C. Is there a second group of leaders, i.e., officials or staff, appointed by individuals or organizations external to the community?
   D. To whom is the staff primarily responsible? Who has the authority to dismiss them? To whom is the staff primarily responsive?
   E. Is there a formal leadership hierarchy with specified levels of authority?
      1. among the resident leaders?
      2. among the staff?
   F. What appear to be the major sources of high social status among the residents?
   G. Does high social status coincide with leadership status?
   H. Are there processes for the prevention of centralization of leadership, power and authority? Is there a policy:
      1. of collective decision making?
      2. of distributing authority among many individuals or subgroups?

V. Central Decision Making
   A. Is there a central decision making structure in the community?
   B. Does this central decision making structure regularly make decisions affecting the total community:
      1. does it allocate space within the community?
      2. does it rule on the acceptance of new members?
      3. does it negotiate contracts in behalf of the community?
   C. Do officials, or resident leaders predominate in the central decision making structure?

1.  is there a single structure in which both types of leaders participate?
2.  are there two separate structures? If so, what is the authority and the domain of each?

D.  If there is no central decision making structure which allocates space, accepts new members, and makes contracts in behalf of the community, how are such decisions made?
   1.  by an organization external to the community?
   2.  by a variety of decentralized decision making systems within the community?
   3.  by all members of the community together?
   4.  central decision making is avoided; there are no important central decisions purposefully made by any individual or group.

VI.  Community Control of Deviant Behavior
   A.  Is there a central mechanism in the community which has the authority to exercise social control?
      1.  under what circumstances does it use coercive power?
      2.  under what circumstances does it use utilitarian power?
      3.  under what circumstances does it use normative power?
   B.  In the absence of, or in addition to the central social control mechanism, how is deviant behavior controlled? What kinds of power are utilized?
      1.  by the residents among each other
         a.  when is normative power (social control) used exclusively?
         b.  when is utilitarian power used?
         c.  when is coercive power used?
      2.  by an external system
   C.  Under what circumstances are individuals or families forced out of the community and by whom?

VII.  The Community and its Subsystems (i.e., families and individuals)
   A.  What are the major subsystems of the community? Are they families, individuals, and/or other kinds of groupings?
   B.  What is the nature of the legal and economic ties between the community, *qua* community, and its subsystems?
   C.  Are there ideological commitments between the community and its members? What is their significance?
   D.  Do individuals view themselves as having important social or moral obligations to the community and its leaders, i.e., they "cannot let them down"?
   E.  Is the idea of community largely coincidental to the individual concerns of the members?
   F.  How are differences between families and the community adjudicated? By whom?

VIII.
    The External Systems
    A.   What are the major external systems to which the community is linked, specifically:
        1.   Economic systems (production, consumption, distribution)
        2.   Political systems (social control)
        3.   Educational systems and communications (socialization)
        4.   Voluntary associations (social participation)
        5.   Systems of social welfare (mutual support)
    B.   What are the principal inputs of each of the external systems into the community?
    C.   Does the community have outputs which contribute to one or more of the external systems? What are they? How essential are they to the external system?
    D.   To what extent do individual members, families, or groups of members create an output which serves the goals of one or more of the external systems?
        1.   Are these outputs made directly to the external system?
        2.   Are these outputs made via the community?
IX.  The Community, its Members, and their Linkage to External Systems
    A.   Which links between the community and the external systems are specifically defined and limited as to function?
    B.   Are there specific agreements concerning the outputs of one system serving as inputs into the other system?
        1.   the community into external systems
        2.   external systems into the community
    C.   Does the community attempt to maximize its autonomy of external systems? Which systems? How?
    D.   Are there direct linkages between external systems and individuals in the community? Are these relationships contractual in nature?
    E.   Are most of the linkages between individuals and external systems via the community, *qua* community?
    F.   Are most of the linkages between individuals and external systems via subsystems of external systems which are located within the community?
    G.   Are most of the linkages between individuals and external systems of an undefined, undeclared, laissez-faire nature?
X.   The Control of External Systems over the Community and its Members
    A.   Are the behavioral norms of members dictated primarily by the external environment or by the community itself?
    B.   Which norms are enforced by utilitarian power, (i.e., primarily by economic means) which derives from outside the community?

C. Which norms are enforced by coercive power which derives from outside the community?

D. To what degree is the community dependent upon external systems for its survival? Which ones? To what degree are these systems benign, or threatening?

XI. Why Individuals and Families Enter the Community

A. Have the members joined voluntarily?

B. Can residents leave whenever they want?

1. temporarily, on what conditions?

2. permanently, on what conditions?

C. How do the reasons for the residents having joined the community relate to the goals of the community?

D. How do the reasons for the residents having joined the community relate to the goals of any of the external systems?

XII. The Linkage of Individuals and Families (horizontally) to Each Other

A. Are there multiple formal links, regulations, or rituals which bind the residents to each other?

B. Are there explicit ideological links among the residents?

C. Are most of the links between residents informal, voluntary, supported by friendship and mutual concern for each other (but *not* by mutual concern for a common goal)?

D. Are links between individuals and among families casual, accidental, and of secondary importance in the lives of the residents?

E. Do families exist within the community in the same sense that they exist in the larger society?

1. What is the role and status of women within the family?

2. What is the role and status of children within the family?

3. What is the role of the head of household?

F. Are there important subsystems within the community which have an important influence upon its organization which have not been mentioned above? What are they?

# SELECTED BIBLIOGRAPHY

Arendt, Hannah. *On Violence.* New York, Harcourt Brace and Co., 1969.

Arnold, Eberhard. *The Early Christians.* Rifton, Plough Publishing Co. (1926) 1970. _____. *Why We Live in Community.* Rifton, Plough Publishing Co. (1927) 1967. _____. *Salt and Light.* Rifton, Plough Publishing Co., 1967. _____. *Love and Marriage in the Spirit.* Rifton, Plough Publishing Co., 1965.

Arnold, Emmy. *Torches Together.* Rifton, Plough Publishing Co., 1964.

Banfield, Edward C. *Politics, Planning, and the Public Interest.* New York, Free Press, 1955.

Becker, Howard. *Social Thought from Lore to Science.* Washington, Harren Press, 1952.

Bennett, John W. *The Hutterian Brethren.* Stanford, Stanford Univ. Press, 1967.

Berger, Peter L. and Luckman, Thomas. *The Social Construction of Reality.* Garden City, Doubleday, 1966.

Blau, Peter M. and Scott, Richard W. *Formal Organizations, A Comparative Approach.* San Francisco, Chandler, 1962.

Broady, Maurice. *Planning for People.* London, Bedford Square Press, 1968.

Buber, Martin. *Paths in Utopia.* Boston, Beacon Press (1949), 1958.

Burgess, Ernest W., editor. *Retirement Villages.* Ann Arbor, Univ. of Michigan, 1961.

Carden, Maren L. *Oneida: Utopian Community to Modern Corporation.* Baltimore, John Hopkins, 1969.

Conkin, Paul K. *Tomorrow a New World.* Ithaca, Cornell University, 1959.

Dahl, Robert. *Who Governs?* New Haven, Yale, 1961.

Dobriner, William. *Class in Suburbia.* Englewood Cliffs, Prentice Hall, 1963.

Etzioni, Amitai. *The Active Society.* New York, Free Press, 1968. _____. "Toward a Theory of Societal Guidance," *American Journal of Sociology,* 1969, 173-187. _____. *Modern Organizations.* Englewood Cliffs, Prentice Hall, 1964. _____. *A Comparative Analysis of Complex Organizations.* New York, Free Press, 1961.

Friedmann, John. "Planning as Innovation - The Chilean Case," *Journal of American Institute of Planners,* Vol. 32, No. 4, July 1966. _____. "Notes on Societal Action," *Journal of American Institute of Planners,* Sept. 1969, 311-318.

Gans, Herbert. *The Levittowners.* New York, Random House, 1967.

Gerth, Hans, and Mills, C. Wright. *From Max*

Gil, David G. "A Systematic Approach to Social Policy Analysis." *Social Service Review* 44:4, 1970. _____. *Unraveling Social Policy* Cambridge, Schenkman Publishing Co. 1973. Weber. New York, Galaxy, 1958.

Goffman, Erving. *Asylums.* New York, Harper, 1961.

Goodman, Paul and Percival. *Communitas.* New York, Random House (1947) 1960.

Gordon, Kermit, ed. *Agenda for a Nation.* Washington, Brookings Institute, 1968.

Gross, Betram M. "A Benevolent Fascism," *Social Policy.* November/December 1970, Vol. I, No. 4.

Haworth, Lawrence. *The Good City.* Bloomington, Indiana Univ. Press, 1963.

Hillery, George A., Jr. *Communal Organizations.* Chicago, Univ. of Chicago, 1968. _____. "Freedom and Social Organization, A Comparative Analysis," *American Sociological Review,* 36:54, 1971. _____. "The Convent, Community, Prison, or Task Force?", *Journal for the Scientific Study of Religion,* VIII: I, 1969.

Homans, George C. *The Human Group.* New York, Harcourt Brace and Co., 1950.

Hunter, Floyd. *Community Power Structure.* Chapel Hill, Univ. of North Carolina, 1953.

Jencks, Edward M.A. *Law and Politics in the Middle Ages.* New York, Holt and Co., 1898.

Jouvenal, Bertrand de. *The Art of Conjecture.* New York, Basic Books, 1967.

Kahn, Alfred. *The Theory and Practice of Social Planning.* New York, Russell Sage, 1969.

Kahn, Herman and Wiener, Anthony J. *The Year 2000.* New York, Macmillan, 1967.

Kanter, Rosabeth M. *Commitment and Community,* Cambridge, Harvard, 1972. _____. "Commitment and Social Organization, A Study of Commitment Mechanisms in Utopian Communities," *American Journal of Sociology,* 30:499-517, 1968.

Kleemeier, Robert W. "Moosehaven, Congregate Living in a Community of the Retired," *American Journal of Sociology,* 59:347-51, 1954. _____. "The Effects of a Work Program on Adjustment Attitudes in an Aged Population," *Journal of Gerontology* 6:373-9, 1951.

Kotler, Milton. *Neighborhood Government.* Indianapolis, Bobbs Merrill, 1969.

Kristol, Irving. "Reflections on Capitalism in a Free Society," *The Public Interest,* Fall 1970.

Langer, Susanne. *Philosophy in a New Key.* New York, New American Library, 1942.

Liell, John T. *Levittown, A Study in Community Planning and Development.* Doctoral Dissertation, Yale University, 1952. Ann Arbor, University Microfilms.

Lindblom, Charles E. *The Intelligence of Democracy.* New York, Free Press, 1965. _____. *A Strategy of Decision.* New York, Free Press, 1963.

Lippitt, Ronald. *The Dynamics of Planned Change.* New York, Harcourt Brace and Co., 1958.

Litwak, Eugene. "A Balance Theory of Coordination Between Bureaucratic Organizations and Community Primary Groups," *Administrative Science Quarterly,* 11:31-58, 1966. _____. "Technological Innovation and Theoretical Functions of Primary Groups and Bureaucratic Structures," *American Journal of Sociology,* 73:471, 1968.

Long, Norton. "The Local Community as a Sociology of Games," *American Journal of Sociology* LXIV:251-261,1958.

Loomis, Charles P. *Social Systems, Essays on their Persistence and Change.* Princeton, Van Nostrand, 1960.

MacIver, R. M. *The Community, A Sociological Study.* London, McMillan, 1917.

Mannheim, Karl. *Ideology and Utopia.* New York, Harcourt Brace, 1969. _____. *Systematic Sociology.* New York, Philosophical Library, 1957.

Marx, Leo. "American Institutions and Ecological Ideals," *Science,* November 27, 1970.

Meyerson, Martin. "Utopian Traditions and the Planning of Cities," *Daedalus.* Winter 1961.

Mumford, Lewis. *The City in History.* New York, Harcourt Brace and Co., 1961.

Nisbet, Robert A. *Social Change and History.* New York, Oxford, 1969. ———. *Community and Power.* New York, Oxford, (1953), 1967.

Oliver, Warner. *Back of the Dream.* The story of the Loyal order of the Moose. New York, Dutton, 1952.

Olson, Mancur, Jr. *The Logic of Collective Action.* New York, Schocken, 1968.

Orzack, Louis H. and Sanders, Irwin T. *A Social Profile of Levittown, New York.* Boston University 1961. Reproduced by University Microfilms, Ann Arbor, Michigan.

Parsons, Talcott and Bales, Robert. *Family, Socialization and Interaction Process.* Glencoe, Free Press, 1955. ———. *The Social System.* Glencoe, Free Press, 1951. ———. *Structure and Process in Modern Society.* Glencoe, Free Press, 1960.

Peattie, Lisa. "Reflections on Advocacy Planning," *Journal of American Institute of Planners, Vol. 34, no. 2,* March 1968.

Pender, David R. *Strategic Hamlets in America.* Columbia, South Carolina, University of South Carolina, 1969.

Rabinowitz, Francine. *City Politics and Planning.* New York, Atherton, 1969.

Ramsoy, Odd. *Social Groups as System and Subsystem.* Glencoe, Free Press, 1962.

Redfield, Robert. *The Little Community.* Chicago, University of Chicago (1956), 1960.

Rein, Martin. "Social Planning: The Search for Legitimacy," in Moynihan, Daniel P., *Toward a National Urban Policy.* New York, Basic Books, 1970.

Reissman, Leonard. *The Urban Process.* Glencoe, Free Press, 1964.

Ross, Murray G. *Case Histories in Community Organization.* New York, Harper, 1958.

Sennett, Richard. *The Uses of Disorder.* New York, Random House, 1970.

Sjoberg, Gideon. "Contradictory Functional Requirements and Social Systems," *Journal of Conflict Resolution* IV, 2, 1960.

Slater, Philip. *The Pursuit of Loneliness.* Boston, Beacon Press, 1970.

Smith, Page. *As A City Upon A Hill,* New York, Knopf, 1966.

Sorokin, Pitrim. *The Crisis of Our Age.* New York, E.P. Dutton, 1941. ———. *Social and Cultural Dynamics.* New York, American Book, 1937.

Stein, Maurice R. *The Eclipse of Community.* New York, Harper, 1960.

Sykes, Gresham M. *The Society of Captives.* Princeton, Princeton Univ. Press, 1958.

Tawney, W. H. *Equality.* London, George Allen & Urwin (1931), 1964.

Theobald, Robert, editor. *Social Policies for America in the 70's.* New York, Doubleday, 1968.

Thompson, James D. *Organizations in Action.* New York, McGraw Hill, 1967.

Toennies, Ferdinand, *Community and Society.* Charles P. Loomis, Transl. New York, Harper, 1957.

Warren, Roland. *The Community in America.* Chicago, Rand McNally, 1963.
———. *Truth, Love and Social Change.* Chicago, Rand McNally, 1971.

Wattel, Harold. "Levittown, A Suburban Community," in Dobriner, William M., *The Suburban Community,* New York, G. P. Putnam, 1958.

Webber, Melvin M. "The Post-City Age," *Daedalus,* Fall, 1968.

Weber, Max. *The Theory of Social and Economic Organization.* Talcott Parsons, Transl. New York, Oxford, 1947.

Wolf, Eric. *Peasants.* Englewood Cliffs, Prentice Hall, 1966.

Wolff, Kurt H. *The Sociology of Georg Simmel.* New York, Free Press, 1950.

Wooton, Barbara. *Freedom Under Planning.* London, George Allen & Unwin, 1945.

Zablocki, Benjamin D. *Christian Because it Works, A Study of Bruderhof Communitarianism,* Doctoral Dissertation, John Hopkins University, 1967, Ann Arbor, University, Microfilms. ———. *The Joyful Community.* Baltimore, Penquin, 1971.

# INDEX

Active generalized cooperation 22, 30, 39, 56, 101-102, 139, 142-143

Age of retrenchment 1, 3-4, 103, 106, 108, 117-118, 129

Alternative society 7-8, 115-118, 120-121, 123, 125-127

Anti-community 7, 25, 29-32, 34-36, 39, 53, 55-57, 70, 91-92, 96, 98, 102, 105, 109, 119, 120-122, 129

Arnold, Eberhard 43, 71-74, 76-77, 79-80, 85, 92

Arnold, Emmy 43, 73

Bader, Scott 144

Bales, Robert F. 155

Banfield, Edward 124

Becker, Howard 1, 129

Bensman, Joseph 102

Berger, Peter L. 114

Blau, Peter 15

Brooks, Richard O. 110-111

Coercive power 11, 16, 22, 30-31, 36-39, 52, 54-55, 66, 68, 82, 84, 115-116, 118, 121, 125-126, 128, 146-148, 156-157

Commercial organization 101

Commitment 14, 100-102

Communal organizations 5-17, 20-23, 25-28, 30, 33-34, 36-37, 39, 55, 84, 88-91, 94-95, 98, 101-102, 104, 108, 111, 113-116, 126-129, 133-152, 154-155, 157

Community:
    types of administered, 25-26, 31-32, 35-36, 38, 41, 51-52, 55, 57, 94, 96, 99, 109-111, 119
    anarchist 119, 122
    crescive 28, 31-32, 35, 57, 70, 96-97, 103, 105, 109-111, 119-120, 122
    designed 25-26, 28, 31-32, 35-38, 41, 66, 70, 94, 96, 99, 109-110, 119
    historical 31, 37
    intentional 25-26, 28, 31-32, 35-36, 41, 56-57, 81, 83-84, 90-92, 94-96, 99, 109-111, 119
    jet set 119, 121
    paradise 119, 122
    planned 31, 38
    Orwell, 1984 25, 29, 32, 119, 122
    solipisistic 25, 29, 30, 32, 70, 96, 119, 121
    total 25, 29-30, 32, 77, 79, 84-85, 91-92, 95-96, 119, 121
    Totalitarianism 25, 29, 30, 32, 39, 53, 55, 57, 96, 119-120, 122
    Traditional village 24, 119, 121, 124-125
    vagabond 119, 122
    voluntary slave 119, 122

Community level 19-20, 23, 26, 29-36, 40, 51-52, 57, 66, 68, 70, 81-82, 84, 86, 88, 91-93, 111, 122

Community relevant elements 20, 26-27, 31

External level 19-21, 23, 26-32, 34-35, 37-38, 51-52, 54, 56-57, 66-67, 69, 82, 84, 86, 96, 98, 122

External linkage 12, 141-142

External systems 17, 27, 38, 53, 68-70, 85, 90, 97, 99, 122, 152, 154

Family level 19-20, 23, 26-32, 40, 43, 51, 57, 66, 69-70, 82, 84, 86, 88, 91-93, 95, 97

Fanon, Frantz 125

Folk village 133, 135, 138, 142, 156

Formal organizations 5, 7-8, 10-17, 20-23, 26-27, 33, 39, 89, 91, 94-95, 100-102, 108-109, 112-115, 117, 127, 133-152, 154-157

Fromm, Erich 122

Geneinschaft 83, 90

Gesellschaft 83, 90

Generalized cooperation 11, 14-15, 17, 22, 30, 33, 36, 82, 86-87, 90, 102, 116, 118, 121, 128, 138-143, 147, 154, 156

Goal 13, 53, 136, 138, (Also see, Specific Goal)

Goal erosion 35-36, 39, 57, 97, 111

Goal displacement 90

Goodman, Paul 3-4

Hawthorne studies 144

Hillery, George A. 9, 83, 89, 94, 98, 103-104, 133-139, 140-141

Homans, George 152-155

Horizontal linkage 12, 14-15, 18, 22, 30, 39, 56, 82, 140-141, 156

Hunter, Floyd 145

Huxley, Aldous 30

Ideological goal 88-92, 95-96

Illich, Ivan 112

Inclusive systems 11-12 , 17, 20, 22, 101, 141-142, 145-146, 148-152, 154, 157

Kahn, Herman 121

Kanter, Rosabeth 83

Kleemeier, Robert W. 46-47

Landecker, Werner 101-102

Langer, Susanne 112 ,138

Levitt, Alfred 58

Levitt, William 59-60, 66, 68

Lewis, Oscar 123

Liell, John 59, 63

Litwak, Eugene 140-141

Long, Norton 145

Luckmann, Thomas 114

Macro systems 11, 108

Maslow, Abraham 112

# INDEX

McCall, Charles 50
McKissick, Floyd 28
Mechanistic interaction 15, 87, 138-139
Mezzo level 97
Micro system 11, 108
Nisbet, Robert 100
Normative elements 114-115
Normative integration 101
Normative power 11, 16, 31, 36-39, 52, 54-55, 66, 68, 82, 84, 115, 118, 126-128, 146-148, 154, 156
Normative structures 12
Official 31, 35, 37-38, 42, 145
Oliver, Warner 42
Olson, Mancur 93-150
Organization 3, 5, 10, 134, 136-137, 140, 145, 151, 155-156
Orientation 13, 53, 133, 136, 151
Orzack, Louis H. 61
Parsons, Talcott 9, 94, 133-136, 148, 155
Partnership 25-26, 28-29, 31, 35-36, 38, 51-54, 57, 66-67, 69, 82, 84, 88, 90-92, 96, 120-121, 151
Passive generalized cooperation 15, 22-23, 26, 37-40, 66, 139-143, 151, 156
Power 15-17, 22, 68, 115, 120, 145-146, 148, 154
Reciprocal goals 26-28, 31, 66, 92
Reciprocal partnerships 25
Reciprocal roles 10
Reich, Charles 123
Reymert, Martin L. 46, 97
Roszak, Theodore 123
Rouse, and Co. 110-111
Sanders, Irwin T. 61
Satyagraha 127
Sennett, Richard 112
Sentimental collectivity orientation 10, 14, 100-102, 149, 151-152

Slater, Philip F. 70, 100, 123
Social system 5-6, 8-12, 14-16, 20-21, 33, 88-89, 91, 95, 97-98, 108, 118, 124, 126, 133-134, 142, 146, 152
Societies, types of,
  folk-sacred society 129
  people-sacred 129
  prescribed sacred 129
Solidary partnership 25-26
Specific goal 13-14, 25, 52-53, 55, 67, 82-83, 88-91, 94-95, 109, 133-137, 141, 144, 147, 149, 151, 156
Spinoza 104
Staff 31, 37-39, 42, 46-48, 50-56, 66-67, 84
Staff-resident split 31, 35, 37-38, 52-53, 56, 66, 70, 82, 84, 97
Status 22, 37, 53-56, 83, 125, 145, 147, 154
Structural free wheeling 11, 15, 36, 87, 104-105, 108, 112, 128, 137-139, 142, 156
Subsystems 10-12, 16-18, 20-22, 24, 29-31, 37, 52-53, 55, 82, 84, 86, 98-99, 133, 135, 137, 140-143, 145, 149-151, 156-157
Total cooperation 29-30, 87
Tugwell, Rexford 4
Utilitarian power 11, 16, 22, 31, 36-39, 52, 54-55, 66, 68, 82, 84, 115, 118, 121, 128, 146-148, 156-157
Vertical linkages 12, 15, 20, 22, 36-39, 87, 141-142, 156
Vidich, Arthur J. 102
Vill 18, 98, 103, 133, 137, 140-143, 145
Warren, Roland 18, 99, 115-116, 141-142
Weber, Max 25
Wolf, Eric 124